T0255139

Lecture Notes in Computer Science 13006

More information about this subseries at http://www.springer.com/series/7412

Suheyla Cetin-Karayumak ·
Daan Christiaens · Matteo Figini ·
Pamela Guevara · Noemi Gyori ·
Vishwesh Nath · Tomasz Pieciak (Eds.)

Computational Diffusion MRI

12th International Workshop, CDMRI 2021
Held in Conjunction with MICCAI 2021
Strasbourg, France, October 1, 2021
Proceedings

Editors
Suheyla Cetin-Karayumak
Harvard Medical School
Boston, MA, USA

Daan Christiaens
KU Leuven
Leuven, Belgium

Matteo Figini
University College London
London, UK

Pamela Guevara
Universidad de Concepción
Concepción, Chile

Noemi Gyori
University College London
London, UK

Vishwesh Nath
Nvidia
Nashville, TN, USA

Tomasz Pieciak
Universidad de Valladolid
Valladolid, Spain

AGH University of Science and Technology
Krakow, Poland

ISSN 0302-9743 ISSN 1611-3349 (electronic)
Lecture Notes in Computer Science
ISBN 978-3-030-87614-2 ISBN 978-3-030-87615-9 (eBook)
https://doi.org/10.1007/978-3-030-87615-9

LNCS Sublibrary: SL6 – Image Processing, Computer Vision, Pattern Recognition, and Graphics

This Springer imprint is published by the registered company Springer Nature Switzerland AG
The registered company address is: Gewerbestrasse 11, 6330 Cham, Switzerland

Preface

In recent years, diffusion MRI (dMRI) has grown to become a central component in neuroimaging and radiology research that now represents 12% of the MRI literature (source: PubMed). In addition, a growing number of clinical applications are being discovered and implemented, including lesion detection, tumour grading, and neurosurgical planning. This expanding interest from researchers and clinicians alike brings about several technical challenges across all stages of the dMRI analysis pipeline: fast and reproducible image acquisition, robust image reconstruction, microstructure modeling, fibre tracking and connectivity mapping, and statistical inference for cross-sectional and longitudinal group studies.

The Computational Diffusion MRI (CDMRI) workshop is organized each year as a satellite event of the Medical Image Computing and Computer Assisted Interventions (MICCAI) conference. For over a decade, CDMRI has provided a platform to share and discuss the latest research and developments in diffusion MRI methods, analysis, and applications. This year, CDMRI took place on October 1, 2021, as a virtual meeting, since ensuing travel restrictions did not permit MICCAI and its satellite events to go ahead in Strasbourg, France, as originally planned.

We received 13 workshop submissions this year, each of which were assessed by at least two members of the Program Committee in a double blind peer review process. Initially, 10 papers were accepted in the first round; the remaining three papers were accepted after revision. These excellent papers are bundled in these proceedings on pages 1–151. In addition, we also had the pleasure to host six invited talks that span technical developments in data acquisition, microstructure modeling and tractography, and cutting-edge applications in cardiac and fetal diffusion MRI.

Finally, this year's CDMRI also hosted the MICCAI Diffusion-Simulated Connectivity Challenge (DiSCo). The objective of this challenge was to evaluate dMRI pipelines for performing quantitative structural connectivity analysis. The challenge specifically targets methods for dMRI pre-processing, estimation of fiber orientation distributions, tractography and streamline filtering algorithms, and algorithms for calculating connectivity strength. Full details about the design and scope of the challenge are provided in the paper by the DiSCo organizers included on page 152 of these proceedings.

October 2021

Suheyla Cetin-Karayumak
Daan Christiaens
Matteo Figini
Pamela Guevara
Noemi Gyori
Vishwesh Nath
Tomasz Pieciak

Organization

CDMRI Organizers

Suheyla Cetin-Karayumak	Harvard Medical School, USA
Daan Christiaens	KU Leuven, Belgium
Matteo Figini	University College London, UK
Pamela Guevara	Universidad de Concepción, Chile
Noemi Gyori	University College London, UK
Vishwesh Nath	Nvidia, USA
Tomasz Pieciak	Universidad de Valladolid, Spain and AGH University of Science and Technology, Poland

Program Committee

Nagesh Adluru	University of Wisconsin–Madison, USA
Maryam Afzali	Cardiff University, UK
Santiago Aja-Fernández	Universidad de Valladolid, Spain
Doru Baran Aydogan	Aalto University, Denmark
Suyash P. Awate	Indian Institute of Technology, Bombay, India
Dogu Baran Aydogan	Aalto University School of Science, Finland
Ryan Cabeen	University of Southern California, USA
Erick Canales-Rodriguez	EPFL, Switzerland
Emmanuel Caruyer	IRISA, France
Alessandro Daducci	University of Verona, Italy
Tom Dela Haije	University of Copenhagen, Denmark
Silvia De Santis	Instituto de Neurociencias de Alicante, Spain
Samuel Deslauriers-Gauthier	Inria, France
Gabriel Girard	EPFL, Switzerland
Jana Hutter	King's College London, UK
Jan Klein	Fraunhofer MEVIS, Germany
Christophe Lenglet	University of Minnesota, USA
Lipeng Ning	Harvard Medical School, USA
Marco Palombo	University College London, UK
Marco Pizzolato	EPFL, Switzerland
Alonso Ramirez-Manzanares	CIMAT, Mexico
Gabriel Ramos-Llordén	Harvard Medical School, USA
Simona Schiavi	University of Genoa, Italy
Thomas Schultz	University of Bonn, Germany
Farshid Sepehrband	University of Southern California, USA

Antonio Tristán-Vega	Universidad de Valladolid, Spain
Pew-Thian Yap	UNC Chapel Hill, USA
Fan Zhang	Harvard Medical School, USA

Invited Speakers

Maryam Afzali	Cardiff University, UK
Santiago Aja-Fernández	Universidad de Valladolid, Spain
Lauren O'Donnell	Harvard Medical School, USA
Jurgen Schneider	University of Leeds, UK
Filip Szczepankiewicz	Lund University, Sweden
Dan Wu	Zhejiang University, China

DiSCo Organizers

Emmanuel Caruyer	IRISA, France
Gabriel Girard	EPFL, Switzerland
Marco Pizzolato	EPFL, Switzerland
Jonathan Rafael-Patiño	EPFL, Switzerland
Jean-Philippe Thiran	EPFL, Switzerland
Raphael Truffet	IRISA, France

Contents

Acquisition

On the b-value Derivation for diffusion-weighted Double-Echo Steady-State (dwDESS) Magnetic Resonance Imaging

Ulrich Katscher[1(✉)], Jakob Meineke[1], Shuo Zhang[2], Björn Steinhorst[1], and Jochen Keupp[1]

[1] Philips Research Europe, Hamburg, Germany
[2] Philips Healthcare, Hamburg, Germany

Abstract. To determine the degree of observed diffusion weighting and to calculate the diagnostically valuable "Apparent Diffusion Coefficient" (ADC) in MR-based "Diffusion-Weighted Imaging" (DWI), a standardized b-value is typically used. To overcome geometrical distortions in standard DWI, diffusion-weighted double-echo steady-state (dwDESS) sequences have been proposed. Balanced bipolar dwDESS, in particular, provides high SNR and is robust against motion. However, a proper definition of the b-value equivalent to standard DWI is not yet well addressed for this bipolar dwDESS. Therefore, this study investigated an estimation for effective b-values in bipolar dwDESS based on approximation of the underlying signal model in the framework of configuration theory. The obtained effective b-values have been validated in phantom as well as *in vivo* breast studies of a healthy human subject, yielding good agreement with the corresponding b-values of standard DWI. Thus, this study enables the quantitative comparison of dwDESS with diffusion-weighted images obtained from different sequences, different measurements, and different scanners, which helps a wide clinical adaption in studies of, for example, cartilage impairments or breast cancer lesions.

Keywords: Double-echo steady-state · Magnetic resonance imaging · Distortion free · Analytic b-value.

1 Introduction

Standard Diffusion-Weighted MRI (DWI) based on Echo-Planar Imaging (EPI) frequently suffers from geometric distortions, severely restricting diagnostic value, e.g. in localization of breast cancer lesions. To overcome this problem, a couple of different approaches have been proposed, e.g., linking DWI with readout-segmented EPI [1], "SPatio-temporal ENcoding" (SPEN) [2], or Double-Echo-Steady-State sequences (dwDESS) [3,4] Among these approaches, dwDESS seems to be particularly promising, however, it is yet unknown how to define a

S. Cetin-Karayumak et al. (Eds.): CDMRI 2021, LNCS 13006, pp. 3–11, 2021.
https://doi.org/10.1007/978-3-030-87615-9_1

corresponding b-value for dwDESS, which would enable a quantitative comparison of dwDESS with other types of DWI. This study investigates a previously introduced signal model [3,4] and derives an effective b-value for a bipolar version of dwDESS [5]. This bipolar version of dwDESS (sequence diagram shown in Fig. 1) is prone to banding artefacts associated to local magnetic field differences, and thus, is not frequently used. To circumvent the issue of banding artefacts, the current study follows the strategy as presented in [5]: the dwDESS sequence is designed in a way to decrease the distance between the banding artefacts below voxel size. Consequently, the bandings within a voxel partly reduce the overall signal intensity but are no longer visible as artefacts. An illustrative phantom example of this strategy is shown in Fig. 2.

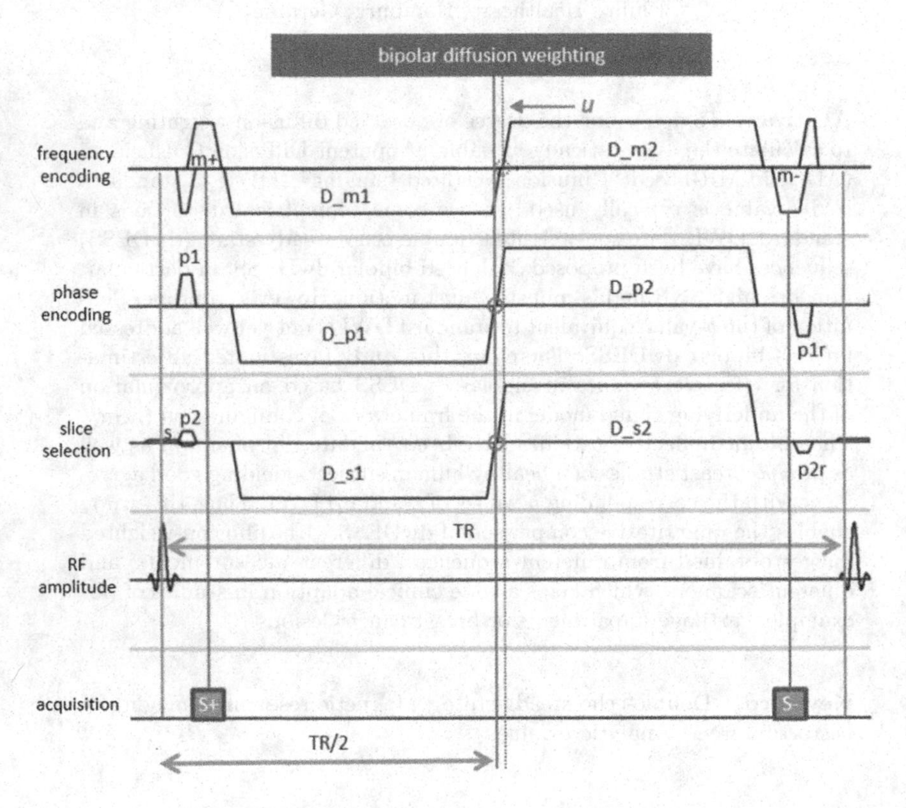

Fig. 1. Sequence diagram of the investigated diffusion weighted Double-Echo Steady-State (dwDESS) sequence with bipolar diffusion gradient. Green boxes indicate the so-called FID S^+(measured shortly after a given RF pulse) and echo S^- (measured shortly before next RF pulse). The diffusion gradient is applied in all three spatial directions (m, p, s) in this sketch for demonstration purpose, but only in m direction in the current study. Please note the small unbalance u (red) of the two diffusion gradient lobes (see Eqs. (3, 4)). (Color figure online)

Fig. 2. Phantom example illustrating how to enable dwDESS with bipolar diffusion gradients. The dwDESS sequence is designed in a way that the distance between the unwanted banding artefacts is decreased (from left to right) until a banding distance below voxel size is no longer visible as artefact.

2 Theory

The general signal model of dwDESS in the framework of configuration theory [3] is summarized in Fig. 3. Based on predefined input quantities (yellow) belonging to tissue (T_1, T_2, diffusivity D) and sequence design (repetition time TR, diffusion gradient duration τ, diffusion gradient amplitude G, flip angle α), the exponential factors E_1 and E_2 are defined (red). These factors lead to several intermediate variables (blue, green), which allow calculation of signal amplitude of FID S^+ measured shortly after a given RF pulse and echo S^- measured shortly before next RF pulse. A double normalization (echo normalized by FID, ratio normalized by fixed gradient amplitude $G = g_0$) yields the desired diffusion signal ΔS (black).

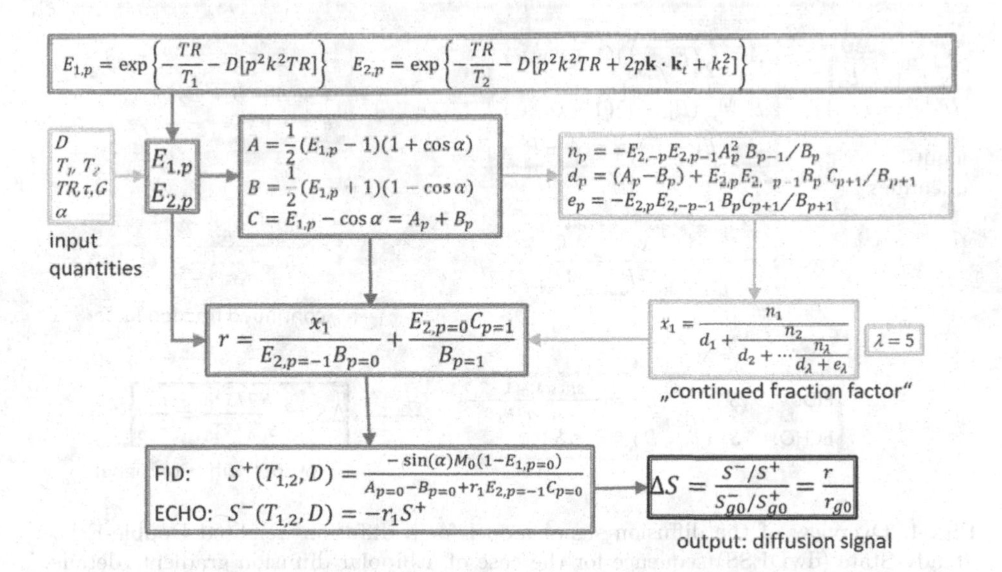

Fig. 3. Overview of the diffusion signal model for diffusion-weighted Double-Echo Steady-State (dwDESS) sequences in the general case [3,4] (details see text).

Figure 4 summarizes the scenario for the specific case of a bipolar diffusion gradient placed symmetrically within TR, consisting of two lobes each with duration $\tau/2$ and opposite polarity [5]. Due to this symmetry, E_1, E_2, and derived quantities simplify (particularly, mode number p completely cancels out), and so does the continued fraction factor x_1, which now can be approximated by first order terms of its numerator and denominator (Fig. 5). Ideally, x_1 comprises an infinite number of levels λ, but usually converges after $\lambda \sim 5$ [3,4]. However, first order terms of numerator/denominator don't change with λ. Thus using these first order terms to approximate x_1, diffusion signal ΔS can be calculated analytically, yielding a mono-exponential function. Analyzing the exponent of this function provides the desired b' which reads

$$b' = \gamma^2 G^2 (3TR^2\tau + \tau^3)/12 \tag{1}$$

and can be used as effective b-value for dwDESS. For the limit of $\tau \Rightarrow$ TR, Eq. (1) reads

$$b' = \gamma^2 G^2 \tau^3/3 \tag{2}$$

and is thus proportional to the textbook definition of b, related to standard DWI sequences. The error introduced by the approximation of x_1 depends on the input variables (G, T_1, T_2, D, TR, α) and is typically below 10% (Fig. 6a).

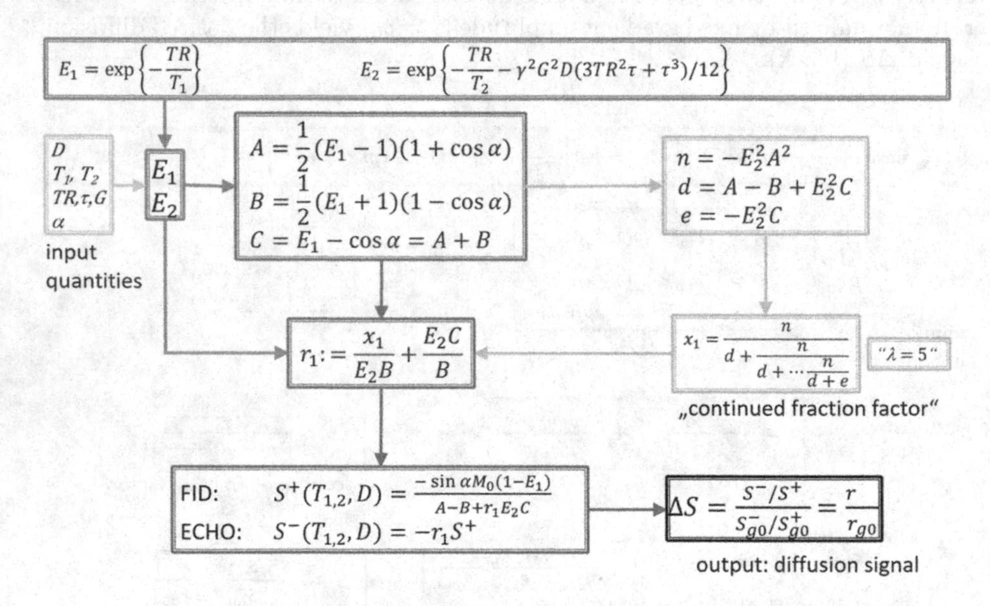

Fig. 4. Overview of the diffusion signal model for a diffusion-weighted Double-Echo Steady-State (dwDESS) sequence for the case of a bipolar diffusion gradient (details see text).

Furthermore, the investigated bipolar version of dwDESS [5] requires a small unbalance u between the two diffusion gradient lobes for suppression of the banding artefacts [6], i.e., two slightly different lobe durations δ and ε

$$x_1 = \frac{n}{d+\frac{n}{d+\frac{n}{y}}} = \frac{dny+n^2}{d^2y+dn+ny} = \frac{-E_2^2A^2y^2+E_2^4(A^4-A^2Cy)}{y^3+2E_2^2(Cy^2-A^2y)+E_2^4(C^2y-A^2C)} \cong \frac{-E_2^2A^2y^2}{y^3} = -\frac{A^2}{A-B}E_2^2$$

Fig. 5. Approximation of continued fraction factor x_1 using first order terms (definition of variables see Fig. 4 and $y \equiv d + e$). The example of $\lambda = 3$ is shown, however, the first order terms included for the approximation (marked) don't depend on λ.

$$\delta = \tau\ (0.5 + u) \quad (3)$$

$$\varepsilon = \tau\ (0.5 - u) \quad (4)$$

In practice, u can be as small as a few percent of the diffusion gradient to sufficiently suppress the bandings [5], thus terms in E_1 and E_2 depending on p are negligible compared to terms not depending on p. The above discussed approximation of b' now yields (with $t \equiv \mathrm{TR}/2 - \tau/2$)

$$b' = \gamma^2 G^2 [t^2(\delta + \varepsilon) + t(\delta^2 + \varepsilon^2) + (\delta^3 + \varepsilon^3)/3 + 2t\delta\varepsilon + \delta^2\varepsilon + \delta\varepsilon^2] \quad (5)$$

reducing to Eq. (1) for $u = 0$.

3 Methods

Two different phantom fluids were investigated, the first having $D = 0.5 \times 10^{-3}$ mm²/s, $T_1 = 420$ ms, $T_2 = 380$ ms, and the second having $D = 1.7 \times 10^{-3}$ mm²/s, $T_1 = 1300$ ms, $T_2 = 610$ ms. Furthermore, the breast of a female volunteer was measured after informed written consent obtained according to local Institutional Review Board, using a commercial 1.5T MRI system (Ingenia, Philips Healthcare, Best, Netherlands). The volunteer was asked to not hold the breath. Two sequences were applied, EPI-DWI (acquired voxel size $1.8 \times 1.8 \times 3.0\,\mathrm{mm}^3$, TR = 997 ms, TE = 62 ms, EPI factor = 87, NSA = 16 yielding roughly 20 s per *b*-value and slice) and dwDESS (acquired voxel size $1.8 \times 1.8 \times 2.0\,\mathrm{mm}^3$, TR = 54 ms, $\mathrm{TE}_1 = 5.4$ ms, $\mathrm{TE}_2 = 48.4$ ms, NSA = 2 yielding roughly 5 s per *b*-value and slice). To suppress banding artefacts, $u = 2.5\%$ was chosen. Diffusion signals were measured with (effective) *b*-values of up to $1200\,\mathrm{s/mm}^2$ for the two phantom fluids and $100/300/500\,\mathrm{s/mm}^2$ for the *in vivo* measurements. Sequences were accelerated by R=3 using compressed sensing. Last, the impact of u on the dwDESS signal was verified with the second phantom fluid (i.e., $D = 1.7 \times 10^{-3}$ mm²/s) by applying the dwDESS sequence using the same parameters as before, but varying u between −10% and +10%, and fixed $b' = 1000\,\mathrm{s/mm}^2$.

(a)

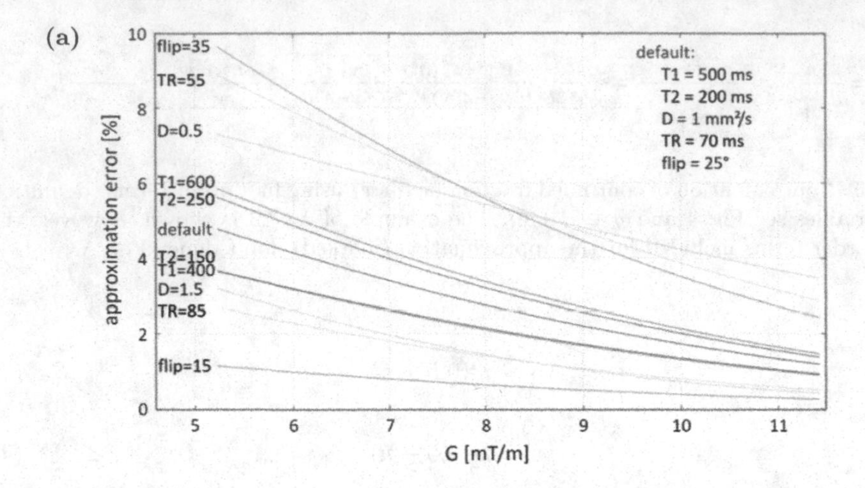

(b)

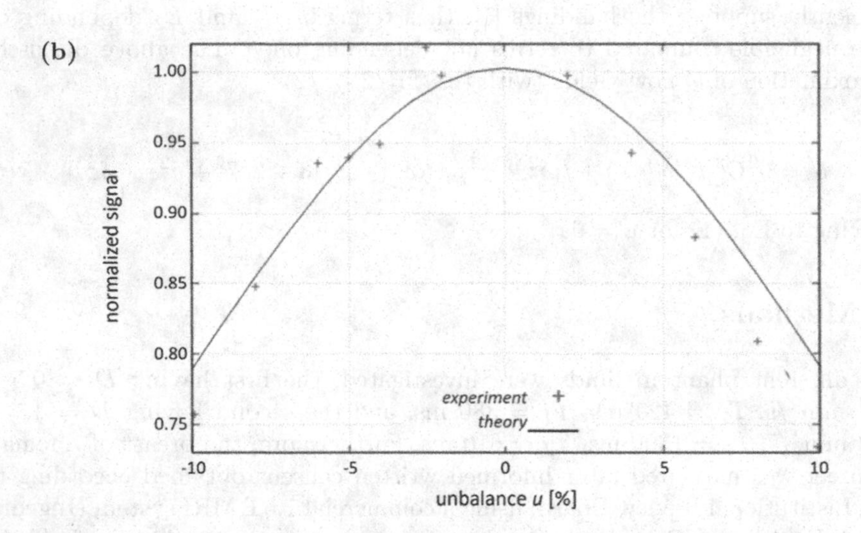

Fig. 6. Impact of applied approximations. (a) Error introduced by approximation of continued fraction factor x_1 versus strength G of diffusion gradient for different exemplary settings of the input variables T_1, T_2, D, TR, and flip angle. (b) Error introduced by unbalance u of diffusion gradient. Since u can be as small as a few percent of the diffusion gradient to sufficiently suppress banding artefacts, the resulting error is expected to be below 5%.

4 Results

The diffusion signals measured for the two phantom fluids with the two diffusion sequences show a high agreement (Fig. 7), based on b for EPI-DWI and b' for dwDESS as described above. Furthermore, the phantom measurements confirmed the signal reduction caused by u as expected from theory (Fig. 6b). Also

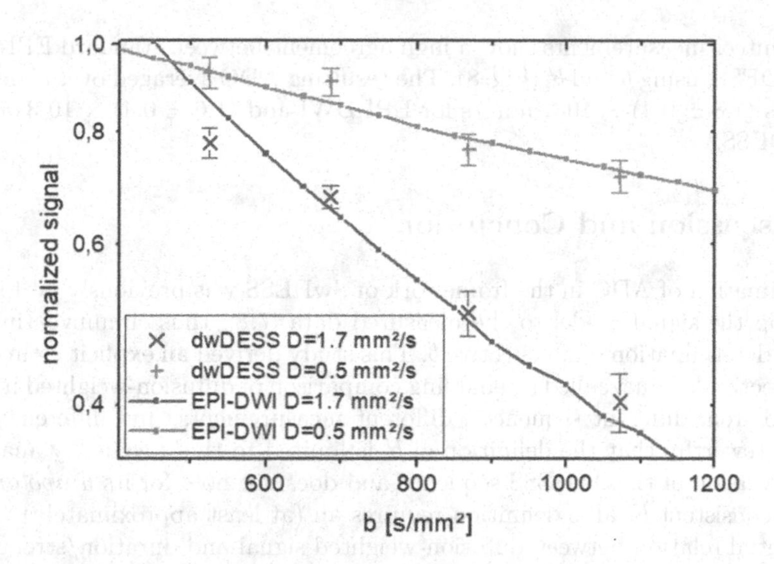

Fig. 7. Results of two phantom fluids with low and high diffusivity (signal scaled logarithmically). For both phantom fluids, a good agreement is obtained between dwDESS with b' as derived in this study and standard EPI-DWI with standard b.

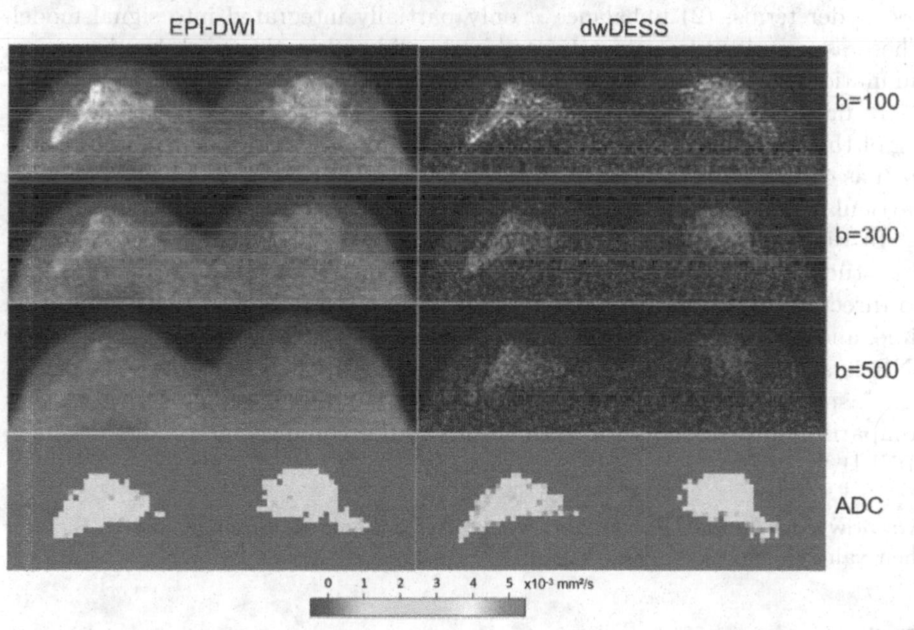

Fig. 8. Comparison of breast DWI of a healthy volunteer. Left column: standard DWI based on EPI, right column: DWI based on DESS (dwDESS). The b-value for dwDESS is calculated using the proposed method and leads to a diffusion weighting and ADC close to standard DWI.

the volunteer measurements show a high agreement between standard EPI-DWI and dwDESS, using b and b' (Fig. 8). The resulting ADC averaged over glandular area was $(1.8 \pm 0.4) \times$ 10-3 mm^2/s for EPI-DWI and $(1.6 \pm 0.4) \times$ 10-3 mm^2/s for dwDESS.

5 Discussion and Conclusion

The estimation of ADC in the framework of dwDESS was previously performed by fitting the signal model to the measured data [7,8], thus circumventing the explicit determination of an effective b. This study derived an explicit estimation of an effective b-value, called b', enabling comparison of diffusion-weighted images obtained from different sequences, different measurements, and different scanners. Noteworthy that the definition of b' is limited to the specifically analyzed *bipolar* version of the dwDESS sequence and does not hold for its *unipolar* version. A consistent b-value definition requires an (at least approximately) mono-exponential relation between diffusion-weighted signal and duration/strength of the diffusion gradient, which has been found only for a bipolar shape and not for an unipolar shape of the diffusion gradient.

The estimation of b' involves two assumptions: (1) approximation of x_1 by first order terms, (2) unbalance u only partially integrated into signal model. These assumptions were investigated and validated in this work by theoretical simulations as well as by experimental studies.

A direct derivation of the b-value in dwDESS helps to gain deeper understanding of the methodology and further pave the way for a wide clinical application, such as cartilage impairment of the knee [7,9] or breast tumors [6,10], which are particularly challenging for standard DWI due to geometric distortions.

In Fig. 8, SNR seems to be higher in EPI-DWI than in dwDESS. However, this study was not designed for SNR comparison between sequences, and thus, no direct SNR ranking is possible due to different voxel size, total acquisition time, and averaging applied in the two sequences. It is expected that comparable SNR is achieved, if equal parameter settings are applied in the two sequences. This aspect shall be investigated in a separate study, together with a general comparison of dwDESS with other means to overcome geometric distortions in DWI [1,2].

Acknowledgements. The authors would like to thank Ch. Stehning and O. Bieri for their valuable support.

References

1. Bogner, W., et al.: Readout-segmented EPI improves the diagnostic performance of diffusion-weighted MR breast examinations at 3.0 T. Radiology **263**(1), 64–76 (2012)
2. Solomon, E., et al.: Overcoming limitations in diffusion-weighted MRI of breast by spatio-temporal encoding. Magnetic Reson. Med. **73**(6), 2163–2173 (2015)

3. Freed, D.E., Scheven, U.M., Zielinski, L.J., Sen, P.N., Hürlimann, M.D.: Steady-state free pre-cession experiments and exact treatment of diffusion in a uniform gradient. J. Chem. Phys. **115**(9), 4249–4258 (2001)
4. Bieri, O., Ganter, C., Scheffler, K.: Quantitative In Vivo diffusion imaging of cartilage using double echo SSFP. Magn. Reson. Med. **68**(3), 720–729 (2012)
5. Keupp, J., Gleich, B., Katscher, U.: Simultaneous acquisition of diffusion weighted images and conductivity maps using a balanced double echo steady state sequence. In: 28th Annual Meeting of International Society of Magnetic Resonance in Medicine, p. 955. Paris (2020)
6. Granlund, K.L., Staroswiecki, E., Alley, M.T., Daniel, B.L., Hargreaves, B.A.: High-resolution, three-dimensional diffusion-weighted breast imaging using DESS. Magn. Reson. Imaging **32**(4), 330–341 (2014)
7. Staroswiecki, E., Granlund, K.L., Alley, M.T., Gold, G.E., Hargreaves, B.A.: Simultaneous estimation of T2 and ADC in human articular cartilage In Vivo with a modified 3D DESS sequence at 3T. Magn. Reson Med. **67**(4), 1086–1096 (2012)
8. Tendler, B.C., Foxley, S., Cottaar, M., Jbabdi, S., Miller, K.L.: Modeling an equivalent b-value in diffusion-weighted SSFP. Magnetic Reson. Med. **84**(2), 873–884 (2020)
9. Miller, K.L., Hargreaves, B.A., Gold, G.E., Pauly, J.M.: Steady-state diffusion-weighted imaging of in vivo knee cartilage. Magn. Reson. Med. **51**(2), 394–398 (2004)
10. Moran, C.J., et al.: Diffusion-weighted double-echo steady-state with a three-dimensional cones trajectory for non-contrast-enhanced breast MRI. J. Magn. Reson. Imaging **53**(5), 1594–1605 (2021)

Quantitative Evaluation of Enhanced Multi-plane Clinical Fetal Diffusion MRI with a Crossing-Fiber Phantom

Hamza Kebiri[1,2](✉), Hélène Lajous[1,2], Yasser Alemán-Gómez[1], Gabriel Girard[3], Erick Canales Rodríguez[3], Sébastien Tourbier[1], Marco Pizzolato[3,4], Jean-Baptiste Ledoux[1,2], Eleonora Fornari[1,2], András Jakab[5,6], and Meritxell Bach Cuadra[1,2,3]

[1] Department of Radiology, Lausanne University Hospital and University of Lausanne, Lausanne, Switzerland
[2] CIBM Center for Biomedical Imaging, Lausanne, Switzerland
[3] Signal Processing Laboratory 5 (LTS5), Ecole Polytechnique Fédérale de Lausanne (EPFL), Lausanne, Switzerland
[4] Department of Applied Mathematics and Computer Science, Technical University of Denmark, Kgs. Lyngby, Denmark
[5] Center for MR Research, University Children's Hospital Zurich, Zurich, Switzerland
[6] Neuroscience Center Zurich, University of Zurich/ETH Zurich, Zurich, Switzerland

Abstract. Diffusion Magnetic Resonance Imaging (dMRI) has become widely used to study *in vivo* white matter tissue properties non-invasively. However, fetal dMRI is greatly limited in Signal-to-Noise ratio and spatial resolution. Due to the uncontrollable fetal motion, echo planar imaging acquisitions often result in highly degraded images, hence the ability to depict precise diffusion MR properties remains unknown. To the best of our knowledge, this is the first study to evaluate diffusion properties in a fetal customized crossing-fiber phantom. We assessed the effect of scanning settings on diffusion quantities in a phantom specifically designed to mimic typical values in the fetal brain. Orthogonal acquisitions based on clinical fetal brain schemes were preprocessed for denoising, bias field inhomogeneity and distortion correction. We estimated the fractional anisotropy (FA) and mean diffusivity (MD) from the diffusion tensor, and the fiber orientations from the fiber orientation distribution function. Quantitative evaluation was carried out on the number of diffusion gradient directions, different orthogonal acquisitions, and enhanced 4D volumes from scattered data interpolation of multiple series. We found out that while MD does not vary with the number of diffusion gradient directions nor the number of orthogonal series, FA is slightly more accurate with more directions. Additionally, errors in all scalar diffusion maps are reduced by using enhanced 4D volumes. Moreover, reduced fiber orientation estimation errors were obtained when used enhanced 4D volumes, but not with more diffusion gradient directions. From these results, we conclude that using enhanced 4D volumes from multiple series should be preferred over using more diffusion gradient directions in clinical fetal dMRI.

S. Cetin-Karayumak et al. (Eds.): CDMRI 2021, LNCS 13006, pp. 12–22, 2021.
https://doi.org/10.1007/978-3-030-87615-9_2

Keywords: Fetal · MRI · Brain · Phantom · Diffusion tensor imaging · Orientation distribution function · Scattered data interpolation

1 Introduction

Diffusion Magnetic Resonance Imaging (dMRI) has been the mainstay of non-invasive white matter investigation *in vivo*. As the diffusion signal is sensitive to the displacement of water molecules in brain tissues, various biophysical models have been proposed for estimating the underlying tissue architecture. These models can either be Gaussian, e.g., diffusion tensor imaging (DTI) is the most simple and widely used model to characterize the diffusion process, or non-Gaussian, e.g., q-ball imaging [31], diffusion spectrum imaging [2,34], and spherical deconvolution [3,30], which estimate Orientation Distribution Functions (ODFs) for resolving multiple intravoxel fiber orientations. However, the unavailability of a ground truth makes the quantitative validation of these models an elusive goal. Monkey brains have been used for connectivity validation of dMRI when compared to histological connectivity obtained from viral tracer injections [1]. Nevertheless, a direct comparison of diffusion orientations at the voxel level is challenging using orientations derived from histological data [27].

On the other hand, phantoms provide an additional possibility for the quantitative evaluation because they offer more controlled, reproducible, and easily accessible experiments. Physical phantoms have been used in dMRI validation setups. For example, the reproducibility of MD measurements was assessed in [17], whereas the recovery of the Ensemble average propagator was validated in a crossing phantom in [23]. In the Fiber Cup [5,10] and ISBI 2018 [26] challenges, tractography reconstructions were compared to ground-truth fiber configurations from physical phantoms. Synthetic software-based phantoms also proved to be a valid alternative to physical phantoms for validation purposes, e.g., see [21,24] and references therein.

In fact, fetal subjects are a sensitive cohort, thus preventing from assessing different acquisition configurations. Hence, the evaluation of our technique on a quantitative dMRI phantom is crucial before applying it to *in vivo* data. However, designing a phantom that matches a fetal brain is extremely complex and challenging. In this work, we use a small size phantom with a customized fractional anisotropy (FA) in the single fiber population in the upper values reported in fetal brains. Indeed, in their atlas, Khan et al. [15] modelled the splenium of the corpus callosum (CC) of a fetus of 37 gestational weeks with an approximately close FA. Similar values were reported both for the genu and the splenium of the CC [8]. Therefore, our phantom is relevant to perform a benchmark analysis in fetuses in the 3rd trimester of gestation. Additionally, the dMRI signal obtained from physical phantoms is similar to *in vivo* data and is more realistic than the dMRI signal obtained from numerical simulations.

Fetal dMRI severely suffers from the unpredictable motion and artifacts caused by the small fetal brain structure that is surrounded by amniotic fluid and maternal organs. Scanning times are typically shorter than that of postnatal studies, limiting the possibility of long diffusion MRI acquisitions based on a large number of diffusion gradient directions and high b-values, which are required to disentangle complex fiber configurations. Furthermore, the use of fast Echo Planar Imaging acquisitions to freeze intra-slice motion leads to highly blurred and distorted images. These images also have a low Signal-to-Noise ratio (SNR), due to the tissue properties of the fetal white matter. Orthogonal scans of anisotropic resolution are usually acquired to overcome these pitfalls. In clinical practice, there is a strong constraint on scanning time, often below 10 min. This does not allow to acquire a high number of orthogonal volumes and a high number of diffusion directions at the same time. Additionally, clinical protocols are not consensual between sites. Typically, clinical fetal brain dMRI have an in-plane resolution of 1–2 mm, a slice thickness of 3–5 mm, the number of gradient directions ranges between 4 and 32, and unique b-values between 400 and 1000 s/mm^2 are employed [12,19,20]. Conversely, in the pioneer research initiative of Developing Human Connectome Project protocol (DHCP) [4,6], up to 141 diffusion volumes can be acquired with multiple b-values (400 and 1000 s/mm^2) and a scanning time of about 15 min per 4D volume.

Our study focuses on the quantitative evaluation of the accuracy of DTI and ODF reconstructions from *in vivo* fetal dMRI acquisitions to identify a good trade-off between the number of series and the number of diffusion gradient directions in a more clinically realistic scenario (summarized in Fig. 1).

2 Methodology

2.1 Materials

Fetal Crossing Phantom - We used a customized fiber crossing phantom (diameter & height of 150 mm) [22] made of two interleaved polyester fiber strands encapsulated in an aqueous solution. The fibers diameter is of 15 µm, the crossing angle between the two strands is approximately 60°, and a customized FA to mimic fetal values in the single fiber population of 0.6 was requested. These values were reported by the vendor who computed them from 128 diffusion-weighted images (DWI) and $b = 1000$ s/mm^2. (Fig. 1A).

MRI Acquisitions - High-resolution (HR) (spatial and angular) images were acquired at 3T (MAGNETOM Prisma-Fit, Siemens Healthcare, Erlangen, Germany), with a 16-channel body array coil and a 32-channel spine coil using a pulsed gradient spin-echo (PGSE) sequence with four different b-values, $400, 700, 1000$ and 3000 s/mm^2. The spatial resolution was 1.5 mm^3 isotropic with a field of view of $256 \times 256 \times 88$ mm^3 , acquired with 61 directions. The echo time (TE) was 52 ms, the repetition time (TR) was 4200 ms and the flip angle was 90°. Only the $b = 700$ s/mm^2 acquisition was considered as the pseudo ground truth (pseudo-GT) in the validation framework.

Low-resolution (LR) acquisitions were performed at 1.5T (MAGNETOM Sola, Siemens Healthcare, Erlangen, Germany), with an 18-channel body array

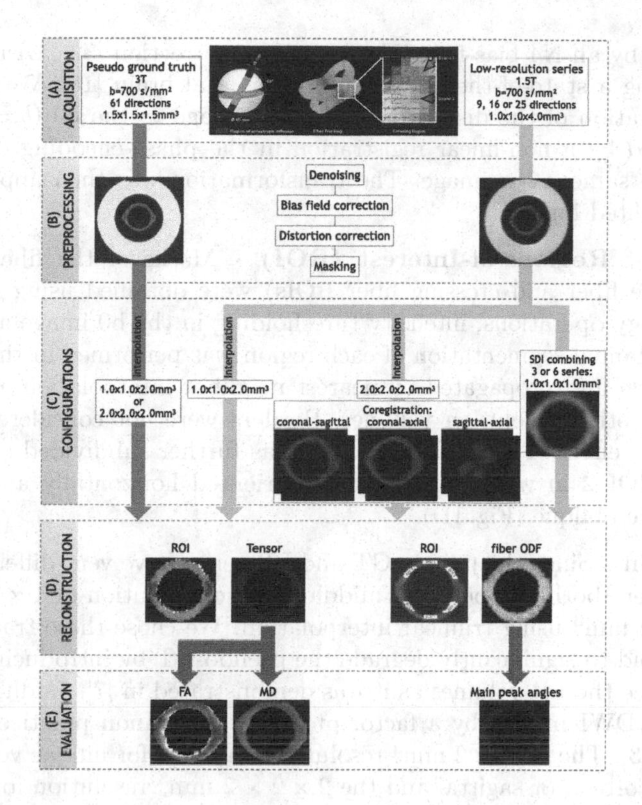

Fig. 1. Summary of our phantom evaluation framework for fetal dMRI acquisitions.

coil and a 32-channel spine coil, using a PGSE sequence (TE = 82 ms, TR = 2000 ms, flip angle = 90°). The acquisition time was approximately one minute per 4D volume. The in-plane resolution was $1 \times 1\,\text{mm}^2$, the slice thickness was 4 mm and the field of view $207 \times 207 \times 69\,\text{mm}^3$. We used $b = 700\,\text{s/mm}^2$ and either 9, 16 or 25 directions, uniformly distributed in the half-sphere. In order to correct non-linear distortions, we also acquired a 1 mm isotropic T2-weighted (T2-w) image using a Sampling Perfection with Application optimized Contrasts using different flip angle Evolution (SPACE) sequence (TE = 380 ms, TR = 3200 ms). Both our data[1] and code[2] will be available to ensure reproducibility of the results.

2.2 Data Processing

Preprocessing - Both the pseudo-GT and LR datasets were preprocessed as follows: a denoising step using a Principal Component Analysis based method

[1] www.zenodo.org/record/5153507#.YQgEA3UzbRY.
[2] www.github.com/Medical-Image-Analysis-Laboratory/FetalBrainDMRI_CrossingPhantom.

[33], followed by an N4 bias-field inhomogeneity correction [32]. Distortion was corrected using a state-of-the-art algorithm for fetal brain [16]. We started by a rigid registration of the distortion free T2-w image to the b0 (b =0 s/mm^2) image followed by a non-linear registration in the phase-encoding direction of the b0 to the same T2-w image. The transformation was then applied to the diffusion-weighted images.

Definition of Regions-of-Interest (ROI) - Masks of the fiber endpoint regions (single fiber and crossing fiber ROIs) were obtained using mathematical morphology operations, intensity thresholding in the b0 image and manual refinement. Manual segmentation of each region was performed in the $1 \times 1 \times 2$ mm^3 resolution and propagated by nearest neighbor interpolation and manual refinement to other resolution volumes. Borders were not considered to avoid partial volume effect. The single fiber ROI was further subdivided in six ROIs: ROI 1 and ROI 2 in which the fibers are oriented horizontally and ROI 3–6 where they are oblique (Fig. 1D).

Interpolation - Since the pseudo-GT and LR series have very different resolutions, they were both mapped to a middle ground resolution of $1 \times 1 \times 2$ mm^3 and $2 \times 2 \times 2$ mm^3 using trilinear interpolation. We chose these trade-off resolutions to avoid to significantly degrade the pseudo-GT by introducing artifacts and to enhance the LR volumes as it was demonstrated in [7]. Additionally, upsampling LR DWI images by a factor of two is a common practice in clinical fetal dMRI [13]. The $1 \times 1 \times 2$ mm^3 resolution was used for unique volumes, i.e., either axial, coronal or sagittal and the $2 \times 2 \times 2$ mm^3 resolution for combined ones, i.e., axial-coronal, axial-sagittal or coronal-sagittal. For the combined volumes, we registered the b0 images of the coronal and the sagittal acquisitions to the b0 image of the axial one using landmarks [9]. This transformation was then applied to the DWI images. To reduce error propagation related to interpolation, we have performed the latter after the preprocessing. We have also computed the different metrics at the different resolutions to quantify variations linked to interpolating the data.

Scattered Data Interpolation - We generated a HR volume from a set of either three or six LR orthogonal series using Scattered Data Interpolation (SDI) reconstruction [25] as implemented in MIALSRTK (version 2.0.1) [28]. It was applied separately to each DWI image and each b0. This consisted in co-registering to an axial reference volume, resampled to isotropic high-resolution, all the series as a first step. Then, the intensity of each voxel in the HR volume grid was computed by averaging the intensities of the corresponding neighboring voxels in the LR volumes using a Gaussian kernel. To match the underlying point spread function of the data, the Gaussian kernel profile was set to be perpendicular to the slice plane with a zero mean and a Full Width at Half Maximum (FWHM of $\sim$2.355 standard deviation) equal to the voxel resolution.

Reconstruction - We reconstructed (1) the diffusion tensor from which we derived both the FA and mean diffusivity (MD) maps and (2) the fiber ODF using the constrained spherical deconvolution (CSD) method [30] from which the main peak (i.e., fiber orientation) was determined. The fiber ODF is represented in the Spherical Harmonics (SH) basis, where an order 4 (15 parameters) was used to best fit all directions (15, 25 for the LR volumes and 61 for the pseudo-GT) and be able to make a one-to-one comparison. In the CSD algorithm, we have constrained the maximum number of peaks to two and the minimum separation angle to 25°. Dipy (version 1.3.0) [11] was used for reconstruction and visualization, and MRtrix3 [29] for fiber ODF visualization.

Evaluation Metrics - To be able to fairly compare diffusion metrics, unbiased by different b-values, we only used as reference the HR data acquired with the same b-value i.e. $b = 700$ s/mm^2 (i.e., pseudo-GT) as the LR data. Scalar maps were evaluated by computing the relative difference between images, i.e., difference between the average LR and the average pseudo-GT map, divided by the average pseudo-GT map. The coefficient of variation (CV, i.e., standard deviation/mean) was also used to quantify the variability of scalar maps.

3 Results

3.1 Scalar Maps

Evaluation of Pseudo-GT - We first assessed the pseudo-GT compared to the diffusion properties given by the vendor. The estimated FA from the pseudo-GT was found to be equal to 0.367 (horizontal orange line in Fig. 2) which did not correspond to the FA reported by the vendor (0.6) in the single fiber population. This is not surprising since FA strongly depends on the acquisition parameters, and in particular on the b-value. Indeed, the same observation was made in the Fiber Cup study [10], where an increase of 75% in the mean FA was reported between $b = 650$ s/mm^2 and $b = 2000$ s/mm^2. The computed FA of 0.367 falls in the same range of FA reported in [14] (using $b = 700$ s/mm^2 & 32 directions) for various fetal brain structures. Conversely, the mean MD = 1.165 mm^2/s was more consistent with the value reported by the vendor (i.e., ~1.2 mm^2/s).

Let us note that scalar maps did not show major differences across different pseudo-GT interpolations, with a CV of 0.5% for single fiber and up to 6.5% for crossing populations. This is lower than the CV of the FA (up to 22%) and MD (up to 12%) values within single and crossing fibers areas of each scalar maps.

Assessment of Enhanced Acquisitions - Figure 2 shows the results from the LR scalar maps for the different configurations compared to the pseudo-GT (two horizontal lines). The orange color refers to the single fiber population and the blue color to the crossing fiber populations, and the bigger the disk diameter the more diffusion directions are used in the reconstruction. For FA, SDI methods outperform the other configurations, especially when considering the single fiber population that shows a difference from the pseudo-GT of 6.1%. Single LR volumes and combinations of pairs are more sensitive to the number of diffusion directions (in these cases, the more directions, the smaller the error), whereas SDI does not show this influence.

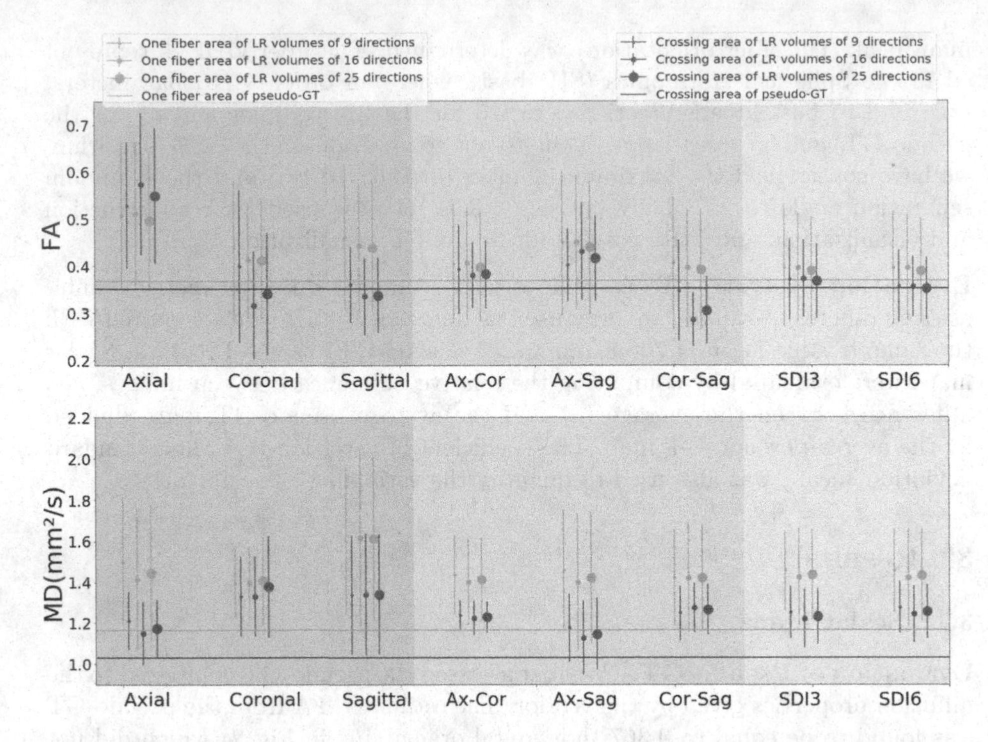

Fig. 2. Scalar maps estimation error compared to the pseudo-GT (horizontal lines, single fiber in orange, crossings in blue, see Fig. 1D). Axial, Coronal, and Sagittal data correspond to single volumes with a resolution of $1 \times 1 \times 2$ mm^3. Ax-Cor, Ax-Sag, and Cor-Sag denote combined volumes with a resolution of 2 mm isotropic. SDI3 and SDI6 are the interpolated scattered data by using three or six 1 mm isotropic volumes, respectively. (Color figure online)

The axial acquisition exhibits a singular behaviour compared to the two other single-volume acquisitions, depicting a higher FA in the crossing area compared to the single fiber area. By inspecting the scanner FA map, we found out an already high FA, particularly in the crossing area of 17% more than in the coronal and the sagittal maps. So merging orthogonal volumes can reduce any potential discrepancy between the different acquisitions (due to outliers and artifacts in the data or due to the anisotropy of the acquisitions capturing the non-symmetrical anatomy across planes of the fibers) and SDI provides the most robust solution.

Differently than FA, MD errors are not influenced by the number of directions neither for single, pairs nor SDI volumes. Both merging pairs of orthogonal volumes and SDI reconstructions help attenuate the high error rate of the sagittal volume by a difference from the average pseudo-GT of about 15% (single fiber population) and 20% (crossing area). The difference between the LR and pseudo-GT values can be explained by the magnetic field strength. Indeed, it was shown in [18], that MD was significantly different between 1.5T and 3T acquisitions.

3.2 Fiber Orientation Errors

Fiber orientations estimation in the pseudo-GT across interpolations is stable in the different ROIs. The maximum standard deviation in ROIs 1–2, where the fibers are close to the x-axis coordinates, is 1.6°, whereas it reaches 4.5° in ROIs 3–6 where the fibers are rotated by around 50°. As depicted in Fig. 3, the angular error of the LR estimated orientations doesn't correlate to the different orthogonal volumes configurations, except for SDI that always shows a lower angular error than, at least, the most under-performing single volume reconstruction. Furthermore, we can observe that the standard deviation of the angular error (vertical lines in Fig. 3) strongly depends on the region of interest. For instance, ROI 1 angles are less variant and closer to the pseudo-GT whereas in ROI 2, the sagittal acquisition compromises the estimated angle of other reconstructions where it belongs. In contrast to ROI 4 and ROI 6, the errors in ROI 3 and ROI 5 are not dramatic as they are located below the mean separation angle of 25°. Importantly, the error difference between the LR and the pseudo-GT volumes is independent of the number of diffusion directions used to compute the main ODF peak.

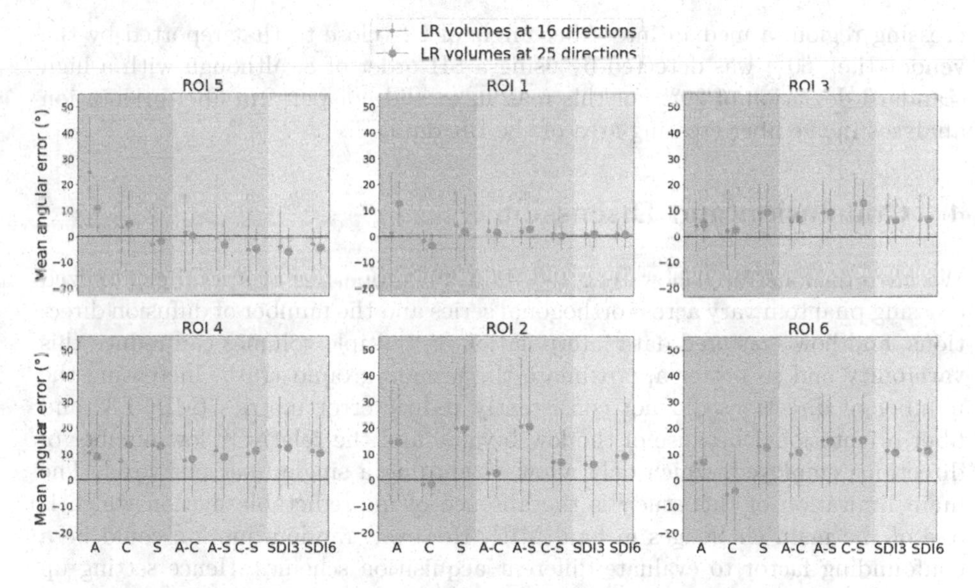

Fig. 3. Mean angular error in different single fiber ROIs corresponding to Fig. 1D for different configurations. Graphs for each ROI are positioned in the corresponding order of their locations on the phantom. A: Axial, C: Coronal, S: Sagittal.

Figure 4A shows fiber ODFs overlaid on the FA map of a LR volume. As can be noted, only very few crossing fibers can be detected at $b = 700$ s/mm^2.

Results shown in Fig. 4B demonstrated that in the HR data, fiber crossings (i.e., two peaks) can only be significantly resolved at $b = 3000$ s/mm^2. In the

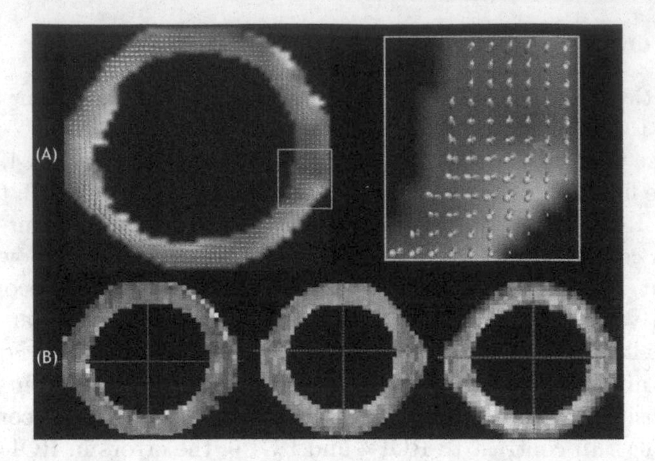

Fig. 4. (A) Fiber ODFs of a LR coronal image overlaid on the FA map. Red arrow: detected crossing. (B) Voxels detected as two peaks in the high resolution acquisition using b-values of 700, 1000 and 3000 s/mm^2 (left to right, respectively). (Color figure online)

crossing region, a median inter-fiber angle of 62° close to that reported by the vendor (i.e., 60°) was detected by using a SH order of 8, although with a high standard deviation of 29°. For this reason, we did not perform fiber orientation analyses in the fiber crossing area of the LR data.

4 Conclusion and Discussion

We have demonstrated how reported diffusion properties of a fetal customized crossing phantom vary across orthogonal series and the number of diffusion directions, and how scattered data interpolation of multiple volumes can reduce this variability and so better approximate the pseudo ground truth. Increasing the number of directions did not consistently reduce error metrics (MD, FA, and fiber orientations) because of the low b-value and the relatively low number of directions employed, which only allow estimating a single fiber per voxel. The main limitation of this study is the absence of unpredictable motion which is one of the main challenges in fetal MRI. However, random motion could be a confounding factor to evaluate different acquisition schemes. Hence setting up a first ideal motion-free scenario to quantify the maximum expected variability of fetal dMRI measurements is a key starting point. Hence, these conclusions have to be taken as an upper bound that can be achieved. In future studies, we plan to extend this work by considering other acquisition protocols (such as the DHCP protocol), by using motion-induced acquisitions for testing different super-resolution reconstruction methods [6], and by implementing scan-rescan analyses in different scanners.

Acknowledgments. This work was supported by the Swiss National Science Foundation (project 205321-182602, grant No 185897: NCCR-SYNAPSY- "The synaptic bases of mental diseases" and the Ambizione grant PZ00P2_185814), the Centre d'Imagerie BioMédicale (CIBM) of the University of Lausanne (UNIL), the Swiss Federal Institute of Technology Lausanne (EPFL), the University of Geneva (UniGe), the Centre Hospitalier Universitaire Vaudois (CHUV), the Hôpitaux Universitaires de Genève (HUG), and the Leenaards and Jeantet Foundations. This work was also supported by the European Union's Horizon 2020 research and innovation programme under the Marie Skłodowska-Curie grant agreement No 754462.

References

1. Aydogan, D.B., et al.: When tractography meets tracer injections: a systematic study of trends and variation sources of diffusion-based connectivity. Brain Struct. Funct. **223**(6), 2841–2858 (2018)
2. Canales-Rodríguez, E.J., Iturria-Medina, Y., Alemán-Gómez, Y., Melie-García, L.: Deconvolution in diffusion spectrum imaging. Neuroimage **50**(1), 136–149 (2010)
3. Canales-Rodríguez, E.J., Legarreta, J.H., Pizzolato, M., Rensonnet, G., Girard, G., Patino, J.R., et al.: Sparse wars: a survey and comparative study of spherical deconvolution algorithms for diffusion MRI. NeuroImage **184**, 140–160 (2019)
4. Christiaens, D., et al.: Fetal diffusion mri acquisition and analysis in the developing human connectome project. In: Proceedings of the Annual Meeting of the International Society of Magnetic Resonance in Medicine (ISMRM) (2019)
5. Côté, M.A., Girard, G., Boré, A., Garyfallidis, E., Houde, J.C., Descoteaux, M.: Tractometer: Towards validation of tractography pipelines. Med. Image Anal. **17**(7), 844–857 (2013)
6. Deprez, M., Price, A., Christiaens, D., Estrin, G.L., Cordero-Grande, L., et al.: Higher order spherical harmonics reconstruction of fetal diffusion mri with intensity correction. IEEE Trans. Med. Imaging **39**(4), 1104–1113 (2019)
7. Dyrby, T.B., et al.: Interpolation of diffusion weighted imaging datasets. NeuroImage **103**, 202–213 (2014)
8. Estrin, G.L., et al.: White and grey matter development in utero assessed using motion-corrected diffusion tensor imaging and its comparison to ex utero measures. Magn. Reson. Mater. Phys. Biol. Med. **32**(4), 473–485 (2019)
9. Fedorov, A., Beichel, R., Kalpathy-Cramer, J., Finet, J., et al.: 3d slicer as an image computing platform for the quantitative imaging network. Magn. Reson. Imaging **30**(9), 1323–1341 (2012)
10. Fillard, P., Descoteaux, M., Goh, A., Gouttard, S., Jeurissen, B., et al.: Quantitative evaluation of 10 tractography algorithms on a realistic diffusion MR phantom. NeuroImage **56**(1), 220–234 (2011)
11. Garyfallidis, E., et al.: Dipy, a library for the analysis of diffusion MRI data. Front. Neuroinf. **8** (2014). https://doi.org/10.3389/fninf.2014.00008
12. Jakab, A., et al.: Disrupted developmental organization of the structural connectome in fetuses with corpus callosum agenesis. Neuroimage **111**, 277–288 (2015)
13. Jakab, A., Tuura, R., Kellenberger, C., Scheer, I.: In utero diffusion tensor imaging of the fetal brain: a reproducibility study. NeuroImage: Clin. **15**, 601–612 (2017)
14. Kasprian, G., et al.: In utero tractography of fetal white matter development. Neuroimage **43**(2), 213–224 (2008)
15. Khan, S., et al.: Fetal brain growth portrayed by a spatiotemporal diffusion tensor MRI atlas computed from in utero images. NeuroImage **185**, 593–608 (2019)

16. Kuklisova-Murgasova, M., et al..: Distortion correction in fetal epi using non-rigid registration with a laplacian constraint. IEEE Trans. Med. Imaging **37**(1) (2017). https://doi.org/10.1109/ISBI.2016.7493522
17. Lavdas, I., Behan, K.C., Papadaki, A., McRobbie, D.W., Aboagye, E.O.: A phantom for diffusion-weighted MRI (DW-MRI). J. Magn. Reson. Imaging **38**(1), 173–179 (2013)
18. Lavdas, I., Miquel, M.E., McRobbie, D.W., Aboagye, E.O.: Comparison between diffusion-weighted MRI (DW-MRI) at 1.5 and 3 tesla: a phantom study. J. Magn. Reson. Imaging **40**(3), 682–690 (2014)
19. Machado-Rivas, F., et al.: Tractography of the cerebellar peduncles in second-and third-trimester fetuses. Am. J. Neuroradiol. **42**(1), 194–200 (2021)
20. Machado-Rivas, F., et al.: Spatiotemporal changes in diffusivity and anisotropy in fetal brain tractography (2021)
21. Maier-Hein, K.H., et al.: The challenge of mapping the human connectome based on diffusion tractography. Nature Commun. **8**(1) (2017)
22. Moussavi-Biugui, A., Stieltjes, B., et al.: Novel spherical phantoms for q-ball imaging under in vivo conditions. Magn. Reson. Med. **65**(1), 190–194 (2011)
23. Ning, L., Laun, F., Gur, Y., DiBella, E.V., Deslauriers-Gauthier, S., Megherbi, T., et al.: Sparse Reconstruction Challenge for diffusion MRI: validation on a physical phantom to determine which acquisition scheme and analysis method to use? Med. Image Anal. **26**(1), 316–331 (2015)
24. Rafael-Patino, J., Romascano, D., Ramirez-Manzanares, A., Canales-Rodríguez, E.J., Girard, G., Thiran, J.P.: Robust Monte-Carlo Simulations in Diffusion-MRI: Effect of the substrate complexity and parameter choice on the reproducibility of results (2019)
25. Rousseau, F., et al.: Registration-based approach for reconstruction of high-resolution in utero fetal MR brain images. Acad. Radiol. **13**(9), 1072–1081 (2006)
26. Schilling, K.G., Gao, Y., Stepniewska, I., Janve, V., Landman, B.A., Anderson, A.W.: Anatomical accuracy of standard-practice tractography algorithms in the motor system - a histological validation in the squirrel monkey brain. Magn. Reson. Imaging **55**, 7–25 (2019)
27. Schilling, K.G., Janve, V., Gao, Y., Stepniewska, I., Landman, B.A., Anderson, A.W.: Histological validation of diffusion MRI fiber orientation distributions and dispersion. NeuroImage **165**, 200–221 (2018)
28. Tourbier, S., Bresson, X., Hagmann, P., Thiran, J.P., Meuli, R., Cuadra, M.B.: An efficient total variation algorithm for super-resolution in fetal brain MRI with adaptive regularization. NeuroImage **118**, 584–597 (2015)
29. Tournier, J.D., Smith, R., Raffelt, D., Tabbara, R., Dhollander, T., et al.: MRtrix3: a fast, flexible and open software framework for medical image processing and visualisation. NeuroImage **202**, 116137 (2019)
30. Tournier, J.D., Yeh, C.H., Calamante, F., Cho, K.H., Connelly, A., Lin, C.P.: Resolving crossing fibres using constrained spherical deconvolution: validation using diffusion-weighted imaging phantom data. Neuroimage **42**(2), 617–625 (2008)
31. Tuch, D.S.: Q-ball imaging. Magn. Reson. Med.: Off. J. Int. Soc. Magn. Reson. Med. **52**(6), 1358–1372 (2004)
32. Tustison, N.J., Avants, B.B., Cook, P.A., Zheng, Y., et al.: N4itk: improved n3 bias correction. IEEE Trans. Med. Imaging **29**(6), 1310–1320 (2010)
33. Veraart, J., Fieremans, E., Novikov, D.S.: Diffusion MRI noise mapping using random matrix theory. Magn. Reson. Med. **76**(5), 1582–1593 (2016)
34. Wedeen, V.J., et al.: Diffusion spectrum magnetic resonance imaging (DSI) tractography of crossing fibers. Neuroimage **41**(4), 1267–1277 (2008)

Microstructure Modelling

A Microstructure Model from Conventional Diffusion MRI of Meningiomas: Impact of Noise and Error Minimization

Letizia Morelli[1(✉)], Giulia Buizza[1], Chiara Paganelli[1], Giulia Riva[2], Giulia Fontana[2], Sara Imparato[2], Alberto Iannalfi[2], Ester Orlandi[2], Marco Palombo[3], and Guido Baroni[1,2]

[1] Department of Electronics, Information and Bioengineering (DEIB), Politecnico di Milano, Milan, Italy
letizia.morelli@mail.polimi.it
[2] National Center of Oncological Hadrontherapy (CNAO), Pavia, Italy
[3] Centre for Medical Image Computing (CMIC), Department of Computer Science, University College London (UCL), London, UK

Abstract. In neuro–oncology microstructural imaging techniques, like diffusion weighted MRI (DW–MRI), have been investigated to non-invasively derive patient–specific parameters that can be used for tumour characterization, treatment personalisation and monitoring, response assessment and prediction of radiotherapy outcomes. In particular, DW–MRI is opening up promising perspectives in radiotherapy applications as it is suitable for characterizing tissues at a microscopic scale (microstructure). However, as advanced MRI is rarely acquired in clinical settings, most studies propose metrics extracted from the conventional apparent diffusion coefficient (ADC), despite it being a sensitive but non–specific metric that encapsulates many features of the underlying tissue.

Starting from conventional ADC, a recently published computational framework showed its potential for tumour characterization at the microscopic scale by means of synthetic cell substrates (which mimic the cellular packing of a tumour tissue) and a simulation tool. The aim of this study was (i) to evaluate the effectiveness of an error correction procedure; (ii) to provide a method that accounts for noise in the computational framework; (iii) to obtain a quantitative description of tumour microstructure from DW–MRI images of meningiomas that helps differentiating patients according to their histological sub–type.

Keywords: DW–MRI · Microstructure · Meningioma

1 Introduction

In the last decade, an increasing number of studies have investigated the potential advantages of biomedical imaging for treatment planning and delivery guid-

S. Cetin-Karayumak et al. (Eds.): CDMRI 2021, LNCS 13006, pp. 25–35, 2021.
https://doi.org/10.1007/978-3-030-87615-9_3

ance in radiation therapy. In particular, quantitative imaging is being investigated to non-invasively derive information on the sources of tumour heterogeneity at multiple spatial scales, which may be extremely relevant for optimizing cancer treatment [2,6,8,14]. Indeed, quantitative information extracted from images can be useful to characterize tumour tissue, e.g., for treatment personalisation or response assessment in radiotherapy, overcoming the limitations of the punctual information obtained from invasive exams like biopsies [3,12,13,15,17,22,23]. In particular, a microscopic characterization (microstructure) derived from diffusion–weighted magnetic resonance imaging (DW–MRI) is showing promising results in various radiotherapy applications [8]. However, despite the great potential, standard protocols are often incompatible with complex DW–MRI sequences, and complex models, such as multi–compartmental [11,16,21] and others [18], that provide information on diffusivity, vascularity, cell density and/or size, are rarely applied in the clinical routine in radiotherapy. For this reason, typically coarser and non–specific indices of tissue microstructure, such as the apparent diffusion coefficient (ADC) [7,19,20], are estimated from conventional DW–MRI acquisitions.

Recently, a computational approach [4] has been proposed to bridge this gap and derive more specific microstructural information from conventional DW–MRI using Monte Carlo simulations of particles diffusion in virtual tumour–like environments. In particular, starting from patients' conventional ADC, this approach enables the estimation of indices of cell size and density, and water apparent diffusivity. However, in its original implementation, the computational approach proposed by Buizza et al. [4] did not explicitly account for experimental noise, which may impact the accuracy and precision of parameters' estimation.

The aim of this study was, therefore, (i) to analyse the added value of an error correction procedure developed to improve the accuracy and precision of parameters' estimation, (ii) to account for noise in the computational framework, thus reducing noise-induced bias, and (iii) to evaluate the ability of the proposed framework to differentiate meningiomas according to their histological sub–type in clinical settings.

2 Materials and Methods

2.1 DW–MRI Acquisition

Twenty–five patients affected by meningioma and enrolled for proton therapy (i.e. an advanced type of radiation therapy which makes use of protons instead of X-rays) at the National Center of Oncological Hadrontherapy (CNAO, Pavia, Italy) were retrospectively selected. The study was approved by the local Research Ethics Board. DW–MRI (b-values = 0,200,400,1000 s/mm^2; resolution = $0.975 \times 0.975 \times 4$ mm; TE/TR = 76/6000 ms; $\alpha = 90°$; slice gap = 0.8 mm; GRAPPA factor = 2; head coil with 32 channels) averaged along three orthogonal directions was acquired on a 3T scanner (Magnetom Verio – Siemens, Erlangen) before treatment. From mono–exponential fits of DW–MRI, ADC maps were computed using three different sets of b-values (b = 200, 1000; b = 400, 1000; b

= 200, 400, 1000 s/mm^2). Gross tumour volumes (GTV) were obtained by rigidly registering post–contrast T1–weighted MRI (VIBE; resolution = 0.6 × 0.6 × 0.6 mm; TE/TR = 2.49/5.35 ms; flip angle = 11°) manual contours on DW–MRI, and manually adjusted according to b0 images, taking care of retaining only the core tumour volume and excluding spurious regions showing, for example, oedema. Information about tumour sub–type (meningothelial, n = 15; atypical, n = 10) were obtained from biopsies.

2.2 Numerical DW–MRI Simulations

DW–MRI signals (N = 3928) were simulated using the Monte Carlo approach implemented in CAMINO [5,10], as reported in a previous study [4], from synthetic substrates. These substrates mimicked the cellular packing of cancerous tissue as an aggregation of ellipsoids with well–defined density and geometrical properties: each synthetic cellular packing is described by fixed and specific microstructural parameters, i.e. cell's radius (R), volume fraction (vf) and diffusivity (D) (range of possible values: R, 2.5–10.0 µm; vf, 0.3–0.6; D, 0.5–3.0 µm^2/ms). Acquisitions parameters for the simulated DW–MRI sequence matched those from patients' acquisitions, and, similarly, ADCs were computed using b = 200, 1000; b = 400, 1000; b = 200, 400, 1000 s/mm^2 to form an ADC space. In this way, each simulated signal was associated with a set of coordinates in the ADC space and specific corresponding R, vf and D.

2.3 From ADC to Microstructure

Following the method in [4], in order to characterize the ADC–microstructure relationship, patient data were mapped in the space of simulated ADCs from synthetic substrates to identify among them the microstructure that best matches the measured data. Specifically, simulated ADCs were partitioned into clusters described by a centroid in the ADC space and by the associated distributions of microstructural parameters. Through an optimization procedure based on the minimization of a LASSO cost function, a set of weights W was obtained to model patients' ADCs as a weighted sum of the clusters' centroids. Since each cluster was described by specific microstructural properties (e.g. median values of R, vf, D), these weights were then used to estimate the microstructural parameters (R, vf, D) coupled to the input data. In particular, for each input data, an estimate of R, vf, and D was obtained as a weighted sum of the median values of the parameters in the clusters, where each weight quantifies the contribution of a cluster to the overall estimate. Additionally, the apparent cellularity (ρ_{app}) was derived as $\rho_{app} = vf/R^3$ to provide a parameter of easier clinical interpretation. Therefore, starting from three ADC values, a median value of R, vf, D and ρ_{app} was estimated.

2.4 Error Correction Procedure

A procedure was implemented to improve the estimates of microstructural parameters by exploiting the absolute estimation error computed from the simulated data, whose true values are known. Such procedure involved: (i) obtaining an initial estimate of microstructural parameters; (ii) calculating the absolute estimation errors (absolute difference between true values and initial parameters estimates) for the simulated data; (iii) considering a set of n first neighboring simulations close to the input data to calculate the mean values of the absolute errors; (iv) adding these average errors to the initial estimates (Fig. 1).

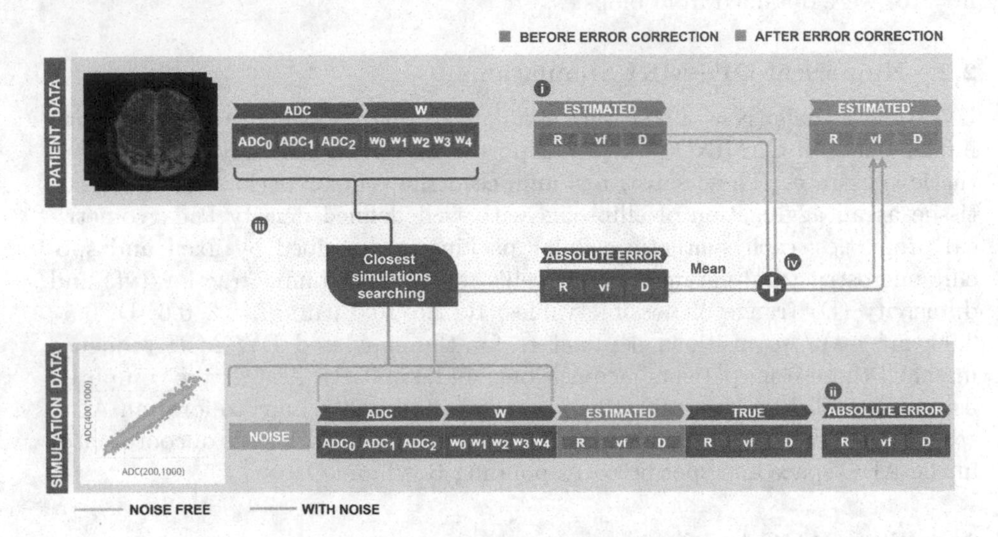

Fig. 1. Schematic representation of the workflow that exploits the absolute estimation error of the method when considering simulated data, to refine the estimation of the microstructural parameters (R, vf, D) coupled to the input data. Parameters estimated before and after error correction are displayed in blue and red, respectively. The green and the orange paths in the workflow indicate the procedure without or with noise added on the simulated signals, respectively. (color figure online)

2.5 Noise

In order to obtain a more representative description of patient data and a more realistic evaluation of the performance of the method, Rician–distributed noise was added to the simulated DW–MRI signals to match the signal–to–noise ratio (SNR) level found in patient data. The SNR from patient data was computed from the b0 image as the ratio between the average signal from a region in the white matter and the standard deviation of the signal from the background [1], and resulted in ≈45.

2.6 Sensitivity and Accuracy Analyses and Application to Patients' Data

Simulated data was initially split in training (80%) and test (20%) sets to evaluate the proposed method with respect to a known ground truth. The overall normalized root mean square error (nRMSE) on a random sample of the simulations (test set) was considered for sensitivity analyses. Such analyses minimized the nRMSE to identify the optimal set of free parameters of the correction procedure, i.e., the number k of clusters (k, candidates from 2 to 10), the number n of closest simulations (n, 13 candidates from 2 to 100) and the type of distance to be used when computing the corrective error (using ADC values only, as dist(ADC), or ADC values and set of weights W, as dist(ADC,W)).

Subsequently, the root mean square error (RMSE), calculated on the test set separately for each parameter, was used to assess the accuracy and precision of the method, before and after the error correction step. The RMSE was then considered to evaluate the behavior of the proposed local error correction procedure when relying on noisy data. Results obtained from 30 different noise realizations were used to compute confidence intervals (confidence = 95%) around the mean RMSE with or without error correction.

Finally, the model, with or without error correction, was applied voxel–wise to patients' GTVs to produce quantitative maps of R, vf, D and ρ_{app}, using all the simulated DW–MRI signals, with or without noise (Fig. 1). Median values were compared between meningothelial and atypical meningioma patients, investigating differences in ADC (200, 400, 1000) and microstructural parameters using a Mann–Whitney U–test ($\alpha = 0.05$).

3 Results

3.1 Noise Free: Sensitivity and Accuracy

Sensitivity analyses for the choice of the free parameters (Fig. 2, in green) showed that, when considering the overall nRMSE, the best result was obtained with $k = 5$ clusters, $n = 4$ close simulations and a distance that considered both ADCs and the set of cluster weights W (with k = 5 and n = 4, overall nRMSE dropped from 0.60 to 0.54 when considering dist(ADC,W) instead of dist(ADC)).

The RMSE calculated, separately for each microstructural parameter, showed that the error correction procedure led to substantially improved accuracy: the RMSE dropped from 2.69, 0.098 and 0.56, to 2.03, 0.076 and 0.20 for R (μm), vf and D (μm^2/ms), respectively (Fig. 3).

3.2 With Noise: Sensitivity and Accuracy

When adding noise to the simulated DW–MRI signals, sensitivity analyses (Fig. 2, in orange) for the definition of free parameters showed that the best result was achieved considering $k = 3$ clusters and $n = 30$ close simulations. The best type of distance was still dist(ADC,W): with $k = 3$ and $n = 30$, the overall nRMSE went from 0.82 to 0.80 when considering dist(ADC,W) instead of dist(ADC).

Accuracy analyses relying on noisy simulated data showed that even if the nRMSE globally increased with respect to the noise–free case, the estimated microstructural parameters benefited from the error correction procedure, each of them showing a lower RMSE than before error correction (Fig. 4). Confidence intervals for the RMSE computed over 30 different noise realizations were 2.71–2.72, 0.10–0.10, 0.56–0.57 before, and 2.57–2.59, 0.10–0.10, 0.50–0.50 after error correction for R (μm), vf and D (μm^2/ms), respectively.

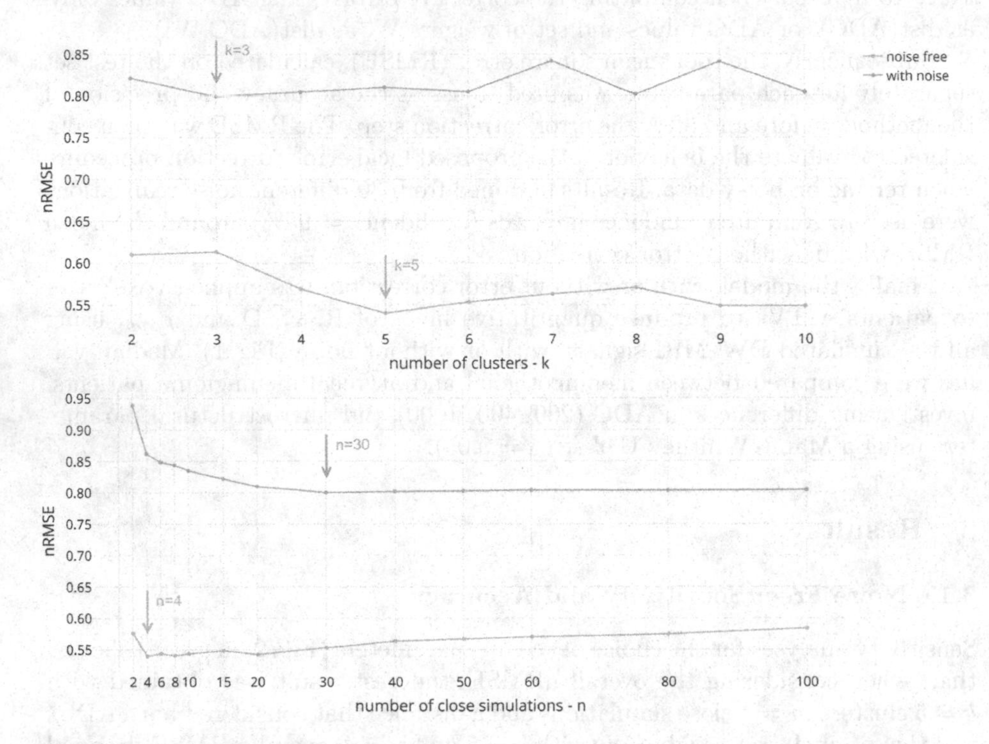

Fig. 2. Sensitivity analyses for the definition of free parameters before (green) and after (orange) adding noise to the simulated DW–MRI signals. The overall nRMSE is shown as a function of the number of clusters k (top row, for each k data is shown for n leading to the optimal nRMSE) or as a function of the number of close simulation n (bottom row, with k = 5 and k = 3 for noise-free and noisy cases, respectively) when considering dist(ADC,W). (Color figure online)

3.3 Patient Data

From the voxel–wise maps (Fig. 5) obtained by applying the error–corrected method with or without noise, it was found that a more significant separation between meningothelial and atypical patients was present when accounting for noise in the simulated signals. Specifically, before applying noise, meningothelial and atypical patients were significantly different only for the median values of ADC ($p = 0.002$ for ADC(200, 400, 1000)), vf ($p = 0.006$) and D ($p = 0.032$) but

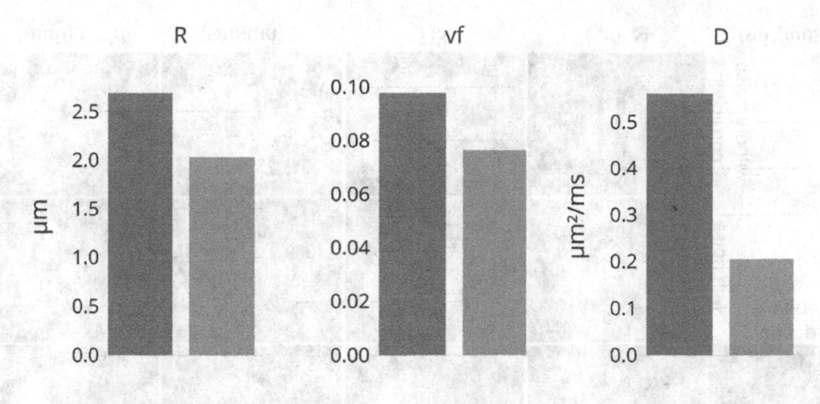

Fig. 3. Comparison of RMSE on R (μm), vf and D (μm^2/ms) obtained for simulations before (blue) and after (red) the correction procedure employing the average absolute estimation error on noise–free simulation data. The RMSE was computed on the test set. (Color figure online)

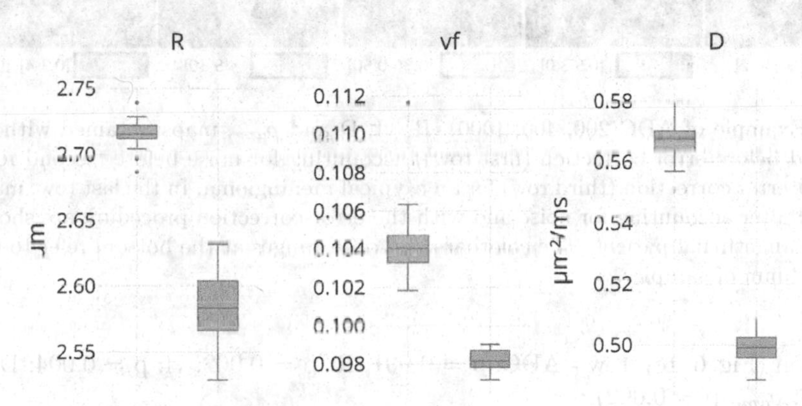

Fig. 4. Boxplots of the RMSE obtained over 30 different realizations of the noise (SNR = 45) before (blue) and after (red) error correction from the test set. (Color figure online)

not for R (p = 0.440) and ρ_{app} (p = 0.190). Instead, after accounting for noise in the simulated signals (Fig. 6, bottom row), all parameters' medians significantly differed between the two groups: ADC (p = 0.001), R (p = 0.009), vf (p = 0.003), D (p = 0.001) and ρ_{app} (p = 0.009). A significant separation between meningothelial and atypical meningiomas was also obtained from the median values of the maps obtained after applying noise to the method without error

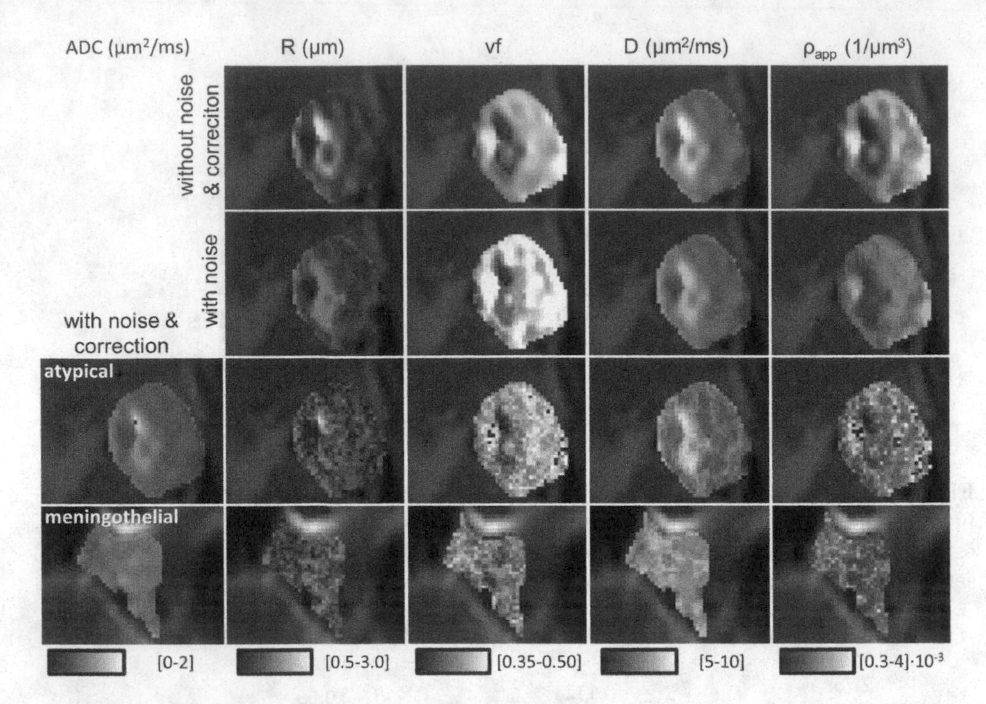

Fig. 5. Example of ADC(200, 400, 1000), R, vf, D and ρ_{app}, maps obtained without noise and before error correction (first row), accounting for noise before (second row) and after error correction (third row) for an atypical meningioma. In the last row, maps obtained after accounting for noise and with the error correction procedure are shown for a meningothelial patient. The colorbars and data ranges at the bottom refer to the whole column of samples.

correction (Fig. 6, top row - ADC: $p = 0.001$; R: $p = 0.002$; vf: $p = 0.004$; D: $p = 0.001$; ρ_{app}: $p = 0.002$).

4 Discussion and Conclusion

In this work, building on top of a previously proposed computational framework [4] aimed at characterizing the ADC–microstructure relationship in terms of predefined tissue properties (R, vf, D, ρ_{app}), an improved procedure to reduce the estimation error of microstructural parameters derived from ADC was introduced. By applying the proposed error correction to simulation data, the RMSE of each estimated property dropped below the resolution at which it was defined (2.5 μm for R, 0.1 for vf and 0.5 μm^2/ms for D), whereas the RMSE from the previously proposed work yielded higher values (with the exception of vf).

Subsequently, Rician noise was added to the simulated DW–MRI signals to match patient data noise levels and the performance of the microstructural model, with or without error correction, was evaluated. Sensitivity analyses showed that the free parameters that yielded optimal performances changed

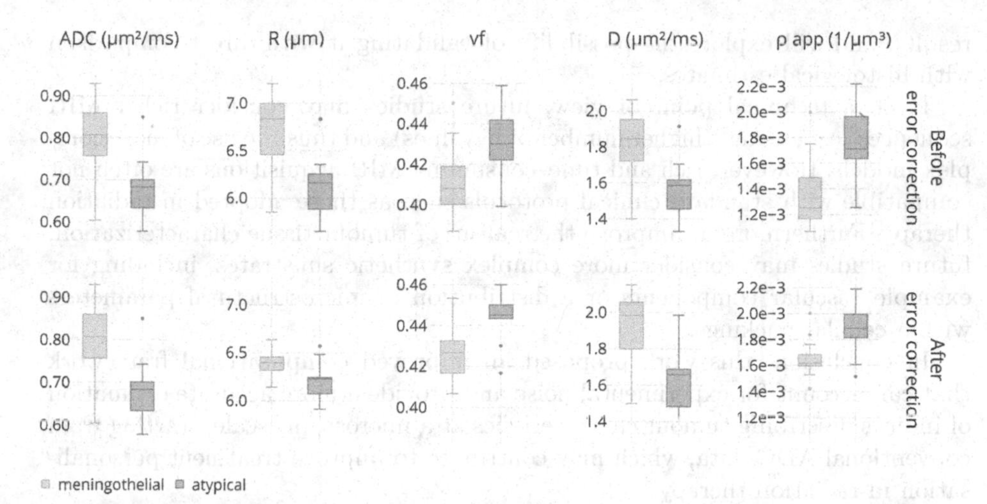

Fig. 6. Boxplots for median values from ADC ($\mu m^2/ms$), R (μm), vf, D ($\mu m^2/ms$) and ρ_{app} (μm^{-3}) for different sub-types obtained by applying the method before (top row) and after error correction (bottom row) with noise on the simulated signals employed for the estimations (SNR = 45). All differences between the two groups are statistically significant ($p < 0.05$).

when feeding noisy simulations. In particular, with noise, the error correction procedure required a higher n of close simulations to better embed the error that the method on average made around the input data. This suggests that it is beneficial to closely match the characteristics of the simulated data to the input data and that, by adding noise, the framework requires a larger neighborhood to properly describe the local error and to correct for any potential bias. This supports very recent findings highlighting the importance of accounting for induced bias in model parameters estimation when supervised machine learning approaches are used for microstructure imaging [9].

When adding noise, the introduced error correction procedure was found to be effective in improving the parameters' estimates. In particular, although the added noise had a lower impact on the accuracy of the method without error correction, it was observed that the error–corrected procedure always showed improved RMSE values.

The potential clinical value of the method was observed in the context of meningioma tumours, as patients with different histological sub–types (atypical and meningothelial) showed distinct tissue properties. Quantitative maps of microstructural parameters derived after error correction showed superior statistical discrimination capabilities when coupling the noise level from patient data to the simulated DW–MRI signals, pointing towards a potentially improved description of microstructural features. As further validation (in histopathological, clinical, and technical terms) of the method is needed, future studies will extend the analysis on a larger cohort of patients, to clinically support the current results, and will explore the possibility of validating it with direct comparison with histological estimates.

From a technical point of view, future studies may consider richer MRI sequences (e.g., using a higher number of b–values) and thus the use of more complex models. However, rich and time–consuming MRI acquisitions are often not compatible with standard clinical protocols such as those adopted in radiation therapy. Furthermore, to improve the realism of tumour tissue characterization, future studies may consider more complex synthetic substrates, including for example vascular components or a distribution of microstructural parameters within cellular packings.

In conclusion, this work proposed an improved computational framework that can account for experimental noise and provide a more accurate estimation of indices describing tumour characteristics at a microscopic scale, starting from conventional ADC data, which may contribute to improve treatment personalisation in radiation therapy.

Acknowledgments. Partially supported by Associazione Italiana per la Ricerca sul Cancro (AIRC), Investigator Grant-IG 2020, project number 24946. MP is supported by UKRI Future Leaders Fellowship MR/T020296/1.

References

1. Aja-Fernández, S., Vegas-Sánchez-Ferrero, G., Tristán-Vega, A.: Noise estimation in parallel MRI: GRAPPA and SENSE. Magn. Reson. Imaging **32**(3), 281–290 2014). https://doi.org/10.1016/j.mri.2013.12.001
2. Bedard, P.L., Hansen, A.R., Ratain, M.J., Siu, L.L.: Tumour heterogeneity in the clinic (2013). https://doi.org/10.1038/nature12627
3. Bontempi, P., et al.: Multicomponent T2 relaxometry reveals early myelin white matter changes induced by proton radiation treatment. Magn. Reson. Med. mrm.28913 (2021). https://doi.org/10.1002/MRM.28913
4. Buizza, G., et al.: Improving the characterization of meningioma microstructure in proton therapy from conventional apparent diffusion coefficient measurements using Monte Carlo simulations of diffusion MRI. Med. Phys. **48**(3), 1250–1261 (2021). https://doi.org/10.1002/mp.14689
5. Cook, P.a., Bai, Y., Seunarine, K.K., Hall, M.G., Parker, G.J., Alexander, D.C.: Camino: open-source diffusion-MRI reconstruction and processing. In: 14th Scientific Meeting of the International Society for Magnetic Resonance in Medicine vol. 14, p. 2759 (2006)
6. Dagogo-Jack, I., Shaw, A.T.: Tumour heterogeneity and resistance to cancer therapies. Nat. Rev. Clin. Oncol. **15**(2), 81–94 (2018). https://doi.org/10.1038/nrclinonc.2017.166
7. Galbán, C.J., Hoff, B.A., Chenevert, T.L., Ross, B.D.: Diffusion MRI in early cancer therapeutic response assessment (2017). https://doi.org/10.1002/nbm.3458
8. Gurney-champion, O.J., et al.: Quantitative imaging for radiotherapy purposes. Radiother. Oncol. **146**, 66–75 (2020). https://doi.org/10.1016/j.radonc.2020.01.026

9. Gyori, N.G., Palombo, M., Clark, C.A., Zhang, H., Alexander, D.C.: Training Data Distribution Significantly Impacts the Estimation of Tissue Microstructure with Machine Learning. bioRxiv p. 2021.04.13.439659 (2021). https://doi.org/10.1101/2021.04.13.439659
10. Hall, M.G., Alexander, D.C.: Convergence and parameter choice for monte-carlo simulations of diffusion MRI. IEEE Trans. Med. Imaging **28**(9), 1354–1364 (2009). https://doi.org/10.1109/TMI.2009.2015756
11. Le Bihan, D.: What can we see with IVIM MRI? NeuroImage **187**, 56–67 (2019). https://doi.org/10.1016/j.neuroimage.2017.12.062
12. Leibfarth, S., Winter, R.M., Lyng, H., Zips, D., Thorwarth, D.: Potentials and challenges of diffusion-weighted magnetic resonance imaging in radiotherapy (2018). https://doi.org/10.1016/j.ctro.2018.09.002
13. Marzi, S., et al.: Early radiation-induced changes evaluated by intravoxel incoherent motion in the major salivary glands. J. Magn. Reson. Imaging **41**(4), 974–982 (2015). https://doi.org/10.1002/jmri.24626
14. Nilsson, M., Englund, E., Szczepankiewicz, F., van Westen, D., Sundgren, P.C.: Imaging brain tumour microstructure (2018). https://doi.org/10.1016/j.neuroimage.2018.04.075
15. Noij, D.P., et al.: Predictive value of diffusion-weighted imaging without and with including contrast-enhanced magnetic resonance imaging in image analysis of head and neck squamous cell carcinoma. Eur. J. Radiol. **84**(1), 108–116 (2015). https://doi.org/10.1016/j.ejrad.2014.10.015
16. Panagiotaki, E., et al.: Noninvasive quantification of solid tumor microstructure using VERDICT MRI. Cancer Res. **74**(7), 1902–1912 (2014). https://doi.org/10.1158/0008-5472.CAN-13-2511
17. Patterson, D.M., Padhani, A.R., Collins, D.J.: Technology Insight: Water diffusion MRI - A potential new biomarker of response to cancer therapy (2008). https://doi.org/10.1038/ncponc1073
18. Reynaud, O.: Time-dependent diffusion MRI in Cancer: tissue modeling and applications. Front. Phys. **5**(November), 1–16 (2017). https://doi.org/10.3389/fphy.2017.00058
19. Surov, A., et al.: Histogram analysis parameters apparent diffusion coefficient for distinguishing high and low-grade meningiomas: a multicenter study. Transl. Oncol. **11**(5), 1074–1079 (2018). https://doi.org/10.1016/j.tranon.2018.06.010
20. Surov, A., Meyer, H.J., Wienke, A.: Correlation between apparent diffusion coefficient (ADC) and cellularity is different in several tumors: a meta-analysis. Oncotarget **8**(35), 59492–59499 (2017). https://doi.org/10.18632/oncotarget.17752
21. Tang, L., Zhou, X.J.: Diffusion MRI of cancer: from low to high b-values. J. Magn. Reson. Imaging **49**(1), 23–40 (2019). https://doi.org/10.1002/jmri.26293
22. Thoeny, H.C., Ross, B.D.: Predicting and monitoring cancer treatment response with diffusion-weighted MRI. J. Magn. Reson. Imaging **16**, 2–16 (2010). https://doi.org/10.1002/jmri.22167
23. Tommasino, F., Nahum, A., Cella, L.: Increasing the power of tumour control and normal tissue complication probability modelling in radiotherapy: recent trends and current issues. Transl. Cancer Res. **6**(5), S807–S821 (2017). https://doi.org/10.21037/tcr.2017.06.03

Generalised Hierarchical Bayesian Microstructure Modelling for Diffusion MRI

Elizabeth Powell[1], Matteo Battocchio[2], Christopher S. Parker[1], and Paddy J. Slator[1](✉)

[1] Centre for Medical Image Computing, Department of Computer Science, University College London, London, UK
p.slator@ucl.ac.uk

[2] Department of Computer Science, University of Verona, Verona, Italy

Abstract. Microstructure imaging combines tailored diffusion MRI acquisition protocols with a mathematical model to give insights into subvoxel tissue features. The model is typically fit voxel-by-voxel to the MRI image with least squares minimisation to give voxelwise maps of parameters relating to microstructural features, such as diffusivities and tissue compartment fractions. However, this fitting approach is susceptible to voxelwise noise, which can lead to erroneous values in parameter maps. Data-driven Bayesian hierarchical modelling defines prior distributions on parameters and learns them from the data, and can hence reduce such noise effects. Bayesian hierarchical modelling has been demonstrated for microstructure imaging with diffusion MRI, but only for a few, relatively simple, models. In this paper, we generalise hierarchical Bayesian modelling to a wide range of multi-compartment microstructural models, and fit the models with a Markov chain Monte Carlo (MCMC) algorithm. We implement our method by utilising Dmipy, a microstructure modelling software package for diffusion MRI data. Our code is available at github.com/PaddySlator/dmipy-bayesian.

Keywords: Bayesian statistics · Bayesian hierarchical model · Microstructure modelling · Diffusion MRI

1 Introduction

Diffusion MRI (dMRI) measures the microscopic motion of water molecules, and is hence sensitive to tissue microstructure. Microstructural modelling combines specifically-designed dMRI acquisitions with a tissue model to enable estimation of parameters relating to tissue microstructure. These techniques have been widely applied in neuroimaging, with prominent examples of brain microstructure imaging including NODDI [10], the standard model of diffusion in neuronal tissue [7] and the spherical mean technique [4]. Microstructural modelling has also provided insights into body MRI [5], for example in prostate cancer [9].

S. Cetin-Karayumak et al. (Eds.): CDMRI 2021, LNCS 13006, pp. 36–47, 2021.
https://doi.org/10.1007/978-3-030-87615-9_4

The core fitting procedure in microstructure imaging estimates model parameters given the observed dMRI signal (Fig. 1, top panel). The vast majority of fitting techniques assume that voxels are independent; in other words, the model is separately fit to the signal in each voxel, usually with nonlinear least squares estimation. An alternative approach is to use an MCMC algorithm to estimate parameter posterior distributions in each voxel [3]. Orton et al. [8] introduced a hierarchical Bayesian model fitting approach for the intravoxel incoherent motion (IVIM [6]) model. Their model breaks the assumption of independent pixels by introducing a Gaussian prior (estimated from the data) over the microstructural model parameters across a region of interest (ROI). By using an MCMC algorithm to fit the Bayesian model, they showed an improvement in IVIM parameter maps of the liver. This approach has also been applied to combined T_2-IVIM modelling in the placenta [2].

In this paper, we generalise the Bayesian approach to apply to any microstructural model, derive the corresponding MCMC algorithm, and implement arbitrary upper and lower parameter bounds. We also utilise regional priors, which may be more appropriate than a global prior for fitting across distinct neurological tissue types. The MCMC algorithm is implemented in Python by utilising the Diffusion Microstructure Imaging in Python (Dmipy [1]) software package. We demonstrate our algorithm on simulations and on HCP data, and show clear advantages over the standard least squares fitting technique.

2 Methods

2.1 General Bayesian Microstructure model

We extend the approach of Orten et al. [8] to a general multi-compartment microstructural model. A schematic of the hierarchical Bayesian framework is shown in Fig. 1.

We consider a general multi-compartment model of N_{comp} compartments, with a set of underlying microstructure-related parameters θ. For notational convenience we group θ by parameter type as

$$\theta = \left\{ \{f_k\}_{k=1}^{N_{comp}-1}, \{x_j\}_{j=1}^{J} \right\} \tag{1}$$

where f_k denotes compartment signal fractions and x_j the other parameters, e.g. diffusivities, orientations, radii, etc. Assuming that relaxation times are fixed across compartments, the signal fractions sum to 1, i.e. $\sum_{k=1}^{N_{comp}} f_k = 1$, meaning that $f_{N_{comp}}$ is not a free parameter but fixed as $1 - \sum_{k=1}^{N_{comp}-1} f_k$.

A general microstructural model comprises a mapping - or *signal equation* - between underlying tissue-related parameters θ and acquisition parameters t_n (typically b-value and gradient direction), and a dMRI signal intensity S_n, i.e.

$$S_n = S_0 g_n(\theta, t_n) \tag{2}$$

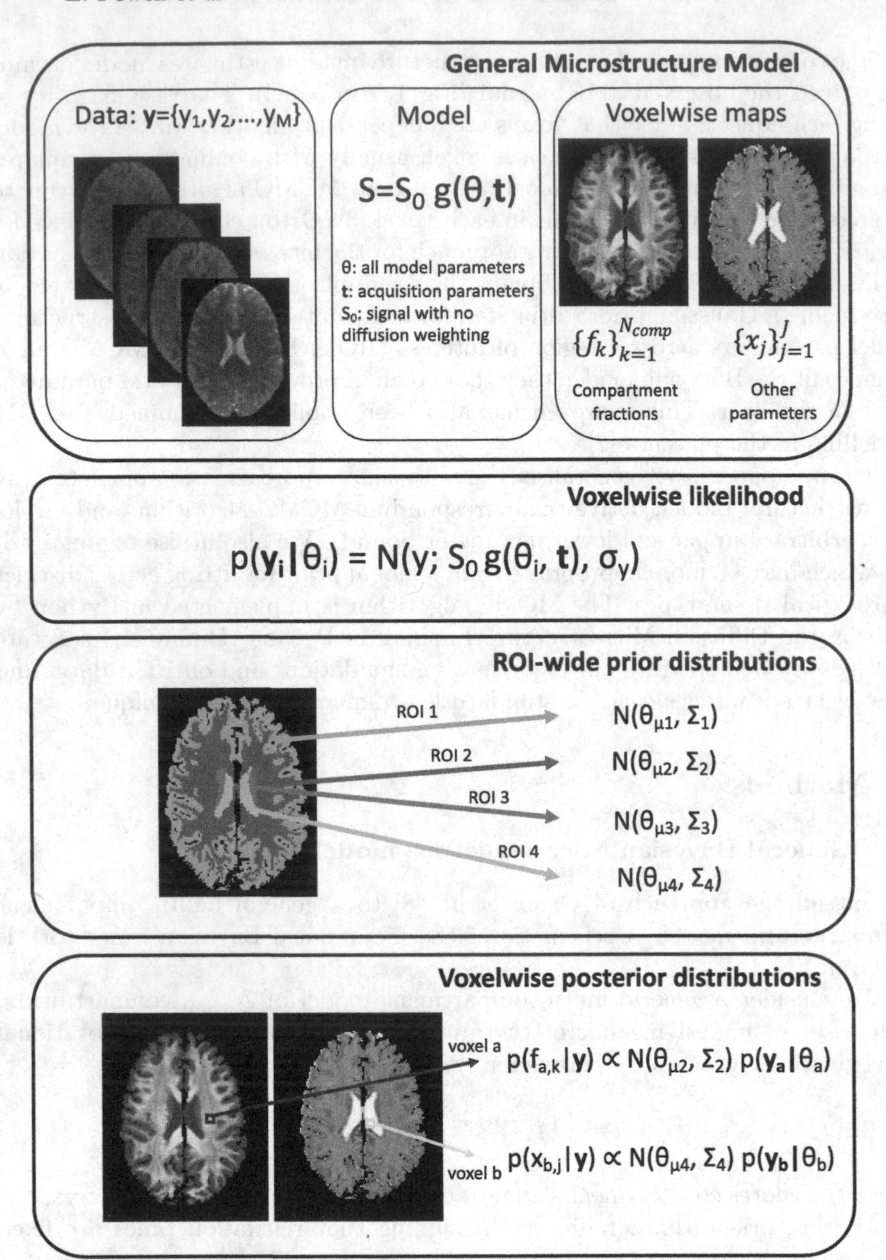

Fig. 1. Schematic of Bayesian hierachical model for a general microstructure model. Top panel defines a general microstructure model, g that maps microstructure-related parameters θ and acquisition parameters $\mathbf{t}$ to dMRI signal S. The microstructure parameters can be grouped by parameter type as $\theta = \left\{ \{f_k\}_{k=1}^{N_{comp}-1}, \{x_j\}_{j=1}^{J} \right\}$ where f_k are the compartment signal fractions and x_j are the other parameters. Second panel defines the voxelwise likelihood function. Third panel displays the ROI-wide Gaussian priors, note that θ_μ and Σ are learnt from the data for all ROIs. Fourth panel displays the voxelwise parameter posterior distributions and corresponding parameter maps.

where S_0 is the signal intensity without diffusion weighting. The experimentally-measured signal in the presence of noise is hence modelled as

$$y_n = S_n + \epsilon_n = S_0 g_n(\theta, t_n) + \epsilon_n \tag{3}$$

where y_n is the measured signal and ϵ_n is noise.

We now consider all measurements for a voxel i - i.e. the signal intensities at all acquisition parameters $\mathbf{t} = \{t_1, ..., t_N\}$ - which we denote $\mathbf{y}_i = [y_1, ..., y_N]^T$. The likelihood, assuming normally distributed noise with variance σ_y^2, is therefore

$$p(\mathbf{y}_i|\theta_i, S_0, \sigma_y^2) = (2\pi\sigma_y^2)^{-N/2} \exp\left(\frac{-1}{2\sigma_y^2}\sum_{n=1}^{N}(y_n - S_0 g_n(\theta_i, t_n))^2\right) \tag{4}$$

where θ_i denotes the microstructural model parameter values in voxel i. Orton et al. [8] demonstrated that the "nuisance parameters" S_0 and σ_y^2 can be marginalised out from Eq. (4) to give the following marginalised likelihood

$$p(\mathbf{y}_i|\theta_i) \propto \left[\mathbf{y}_i^T\mathbf{y}_i - (\mathbf{y}_i^T\mathbf{g}_i)^2/\mathbf{g}_i^T\mathbf{g}_i\right]^{-N/2} \tag{5}$$

where $\mathbf{g}_i = [g_1(\theta_i, t_1), ..., g_N(\theta_i, t_N)]$ are the model predicted signals for voxel i.

2.2 Parameter Transforms

Microstructure model fitting needs to enforce physically reasonable minimum and maximum values of parameters; for example, diffusivities need to be positive. Here we generalise the transforms used by Orton et al. [8] to enable arbitrary minimum and maximum constraints. For a parameter p, we define a transform

$$p' = \log(p - p_{min}) - \log(p_{max} - p). \tag{6}$$

which maps the interval (p_{min}, p_{max}) to $\mathbb{R}$. By defining the Bayesian prior on the transformed parameter p', we therefore constrain p between p_{min} and p_{max}. Default values for p_{min} and p_{max} are set as the minimum and maximum values defined in the Dmipy variable "model.parameter_ranges"; however, they can also be manually defined by the user.

2.3 Bayesian Shrinkage Priors

Orton et al. [8] used a multivariate Gaussian Bayesian shrinkage prior on the IVIM model parameters, with the prior defined over a single user-defined ROI. The Bayesian fitting method is generalised here to the multiple ROI case simply by running the derived MCMC algorithm separately on the voxelwise dMRI data from each ROI; however, note that here and throughout the methods section we consider the single ROI case for brevity, without loss of generality. The prior generalised for any microstructural model is denoted

$$p(\theta|\theta_\mu, \Sigma) = N(\theta; \theta_\mu, \Sigma) \tag{7}$$

where θ_μ is a vector whose elements encode the prior means of the parameters, Σ is their covariance and $N(\theta;\theta_\mu,\Sigma)$ denotes the multivariate normal probability density function (PDF) with variable θ, mean θ_μ and covariance Σ. Again, we emphasise that θ_μ and Σ are estimated from the data.

To generalise from Orton et al.'s [8] two-compartment model to an arbitrary multi-compartment model, all signal fractions must to sum to one. We enforce this (following Harms et al. [3]) by modifying the prior to

$$p(\theta_i|\theta_\mu,\Sigma) = \begin{cases} N(\theta_i;\theta_\mu,\Sigma) & \text{if } \sum_{j=1}^{N_{comp}-1} f_j \leq 1 \\ 0 & \text{otherwise} \end{cases} \tag{8}$$

To complete the model we define a hyper-prior on θ_μ and Σ as a non-informative Jeffrey's prior

$$p(\theta_\mu,\Sigma) = |\Sigma|^{-1/2} \tag{9}$$

2.4 Posterior Distributions

Each ROI has its own posterior distribution, which can be written as [8]

$$p(\theta_{1:M},\theta_\mu,\Sigma|\mathbf{y}_{1:M}) \propto p(\mathbf{y}_{1:M}|\theta_{1:M})p(\theta_{1:M}|\theta_\mu,\Sigma)p(\theta_\mu,\Sigma)$$

where $\theta_{1:M} = \{\theta_1,\theta_2,...,\theta_M\}$ are the parameters and $\mathbf{y}_{1:M} = \{\mathbf{y}_1,...,\mathbf{y}_M\}$ the dMRI data for all voxels in the ROI. Substituting in equations (5), (7), (9) gives

$$p(\theta_{1:M},\theta_\mu,\Sigma|\mathbf{y}_{1:M}) \propto \left(\prod_{i=1}^{M}\left[\mathbf{y_i}^T\mathbf{y_i} - (\mathbf{y_i}^T\mathbf{g_i})^2/\mathbf{g_i}^T\mathbf{g_i}\right]^{-N/2}\right)\left(\prod_{i=1}^{M}N(\theta_i;\theta_\mu,\mathbf{\Sigma})\right)|\Sigma|^{-1/2}$$

from which we can draw samples with an MCMC algorithm.

2.5 MCMC algorithm

The MCMC algorithm is derived here, and given as pseudocode in Algorithm 1.

ROI-wide Parameters. Following Orton et al. [8], the MCMC updates for the ROI-wide prior parameters θ_μ and Σ are Gibbs moves. The conditional distributions are (up to proportionality)

$$p(\theta_\mu|\theta_{1:M},\Sigma,\mathbf{y}_{1:M}) \propto \prod_{i=1}^{M} N(\theta_i;\theta_\mu,\mathbf{\Sigma}) = N(\theta_\mu;m,V)$$

where $m = M^{-1}\sum_{i=1}^{M}\theta_i$ $V = M^{-1}\Sigma$, and the second line comes from rearranging the multivariate normal PDF so that θ_μ is the variable. The MCMC update is therefore sampled as follows

$$\theta_\mu \sim N(m,V) \tag{10}$$

where $N(m, V)$ is a multivariate normal distribution with mean m and covariance V. Following the same steps for Σ (see Orton et al. [8] for full details) gives the MCMC update for Σ

$$\Sigma \sim W^{-1}(\Phi, M-3) \tag{11}$$

where $\Phi = \sum_{i=1}^{M}(\theta_i - \theta_m u)(\theta_i - \theta_m u)^T$ and W^{-1} is the inverse-Wishart distribution.

Voxelwise Parameters. For the non-signal fraction voxelwise parameters the posterior distribution up to proportionality is

$$p(x_{i,j}|x_{i,-j}, \{f_{i,k}\}_{k=1}^{N_{comp}-1}, \theta_\mu, \Sigma) \propto \left[\mathbf{y_i}^T\mathbf{y_i} - (\mathbf{y_i}^T\mathbf{g_i})^2/\mathbf{g_i}^T\mathbf{g_i}\right]^{-N/2} N(\theta_i; \theta_\mu, \Sigma) \tag{12}$$

where $x_{i,j}$ is the value of parameter x_j in voxel i, $x_{i,-j} = \{x_{i,1}, ..., x_{i,j-1}, x_{i,j+1}, ..., x_{i,J}\}$ denotes the set of all non-signal fraction parameters except $x_{i,j}$, and $\{f_{i,k}\}_{k=1}^{N_{comp}-1}$ are the signal fractions for voxel i.

As in Orton et al. [8], we sample from this with a Metropolis-Hastings algorithm. Proposed parameters are first sampled from Gaussian distributions as

$$x_{i,j}^* \sim N(x_{i,j}, w_{x_{i,j}}) \tag{13}$$

where $x_{i,j}$ is the current value of the parameter, $x_{i,j}^*$ is the proposed parameter value and $w_{x_{i,j}}$ is the variance of the proposal distribution, which should reflect the scale of the parameter and can be tuned for optimal algorithm performance.

The acceptance probability utilises the ratio of the posterior distributions for $x_{i,j}$ and $x_{i,j}^*$

$$\alpha(x_{i,j} \to x_{i,j}^*) = \min\left\{1, \frac{p(x_{i,j}^*|x_{i,-j}, \{f_k\}_{k=1}^{N_{comp}-1}, \theta_\mu, \Sigma)}{p(x_{i,j}|x_{i,-j}, \{f_k\}_{k=1}^{N_{comp}-1}, \theta_\mu, \Sigma)}\right\} \tag{14}$$

where the values on the right of the posterior are the current parameter values in the MCMC algorithm.

The signal fraction parameter MCMC moves are the same, except that the posterior distributions now contain the terms enforcing $\sum_{k=1}^{N_{comp}} f_k = 1$, i.e.

$$p(f_{i,k}|x_{i,1}, ..., x_{i,J}, f_{i,-k}, \theta_\mu, \Sigma) \propto \begin{cases} \left[\mathbf{y_i}^T\mathbf{y_i} - (\mathbf{y_i}^T\mathbf{g_i})^2/\mathbf{g_i}^T\mathbf{g_i}\right]^{-N/2} & \text{if } \sum_{k=1}^{N_{comp}-1} f_k \leq 1 \\ 0 & \text{otherwise} \end{cases} \tag{15}$$

where $f_{i,-k} = \{f_{i,1}, ..., f_{i,k-1}, f_{i,k+1}, f_{i,K}\}$ are the other signal fractions apart from $f_{i,k}$. Again we sample proposed values as

$$f_{i,k}^* \sim N(f_{i,k}, w_{f_{i,k}}) \tag{16}$$

where $f_{i,k}$ is the current signal fraction. The acceptance probabilities are

$$\alpha(f_{i,k} \to f_{i,k}^*) = \min\left\{1, \frac{p(f_{i,k}^*|x_{i,1}, ..., x_{i,J}, f_{i,-k}, \theta_\mu, \Sigma)}{p(f_{i,k}|x_{i,1}, ..., x_{i,J}, f_{i,-k}, \theta_\mu, \Sigma)}\right\} \tag{17}$$

Metropolis-Hastings Acceptance Ratio. We tuned the Metropolis-Hastings jumping variances w_{θ_i} during the burn-in period to achieve an acceptance ratio that samples the posterior distribution efficiently. Following Orton et al. [8], at every 100 MCMC steps we applied the update rule

$$w_{\theta_i} = w_{\theta_i} 101/\left(2\left(101 - R_{\theta_i}\right)\right) \tag{18}$$

where R_{θ_i} is the number of times the proposed parameter update was accepted in the previous 100 steps. This scheme aims to adjust the jumping variances such that an acceptance rate of approximately 50% is achieved.

2.6 Models

The MCMC algorithm was tested using the ball-stick model, defined as

$$g(\theta, t) = f_{par} \exp\left(-bD_{par}(\mathbf{n}.\mathbf{g})\right) + (1 - f_{par}) \exp\left(-bD_{iso}\right) \tag{19}$$

where b is the b-value, $\mathbf{g}$ is the gradient direction and $\mathbf{n}$ is the stick orientation, which is parameterised by two angles ϕ_1 and ϕ_2 constrained such that $\phi_1 \in (0, \pi)$ and $\phi_2 \in (-\pi, \pi)$. The signal fractions were constrained as $f_k \in (0.01 - 0.99)$ and the diffusivities as $D_{par}, D_{iso} \in (0.1 - 3)\,\mu\text{m}^2/\text{ms}$.

2.7 Algorithm Implementation

Note that while all distributions have been presented in the linear scale, they were calculated in log-scale for numerical convenience. Parameter values were initialised with a voxelwise least squares fit, estimated using the Dmipy brute2fine option [1]. The MCMC algorithm was then run for 2000 steps with a burn-in of 1000 steps; weights were updated every 100 steps during the first half of the burn-in period. In our experience this was sufficient to sample the posterior distributions, and aligns with the work of Harms et al. [3]. We calculated model parameter posterior distributions and representative statistics from the 1000 MCMC samples after the burn-in. Parameter maps were generated using the mean of the posterior distributions in each voxel.

2.8 Data

To test the MCMC algorithm's ability to infer correct model parameter values, we ran simulations using the Shepp-Logan phantom. We generated synthetic images with a matrix size of 128 × 128 and defined ground truth parameters in each major region (see Fig. 3A, top row). We then simulated the signal in each voxel using Dmipy's simulate_signal function with the same acquisition parameters as the Human Connectom Project (HCP) data (see below), added Gaussian noise to give a signal-to-noise ratio (SNR) of 10 in the $b = 0$ data, and ran the MCMC algorithm on these synthetic datasets. We perturbed initial parameter values to verify that the algorithm could find the global minimum.

Algorithm 1. MCMC algorithm for Bayesian model fitting of a general microstructural model. This pseudocode describes the algorithm for a single ROI. For multiple ROIs, the algorithm is simply run separately on each ROI.

for voxels $i = 1$ to $i = N$ **do**
  Calculate initial values for voxelwise parameters: $\{x_{i,j}\}_{j=1}^{J}$, $\{f_{i,k}\}_{k=1}^{N_{comp}}$ with least squares estimation
end for
$S \leftarrow$ number of MCMC steps
for MCMC steps $s = 1$ to $s = S$ **do**
  $\theta_{\mu}^{(s)} \leftarrow$ sample from Equation (10)
  $\Sigma^{(s)} \leftarrow$ sample from equation (11)
  for voxels $i = 1$ to $i = N$ **do**
    for non-signal fraction parameters $j = 1$ to $j = J$ **do**
      $\epsilon_{x_{i,j}} \leftarrow N(x_{i,j}^{(s-1)}, w_{x_{i,j}})$
      $x_{i,j}^{*} \leftarrow x_{i,j}^{(s-1)} + \epsilon_{x_{i,j}}$
      calculate $\alpha(x_{i,j} \rightarrow x_{i,j}^{*})$ from Equation (14)
      $u \leftarrow$ sample from unif$(0, 1)$
      if $u < \alpha$ **then**
        $x_{i,j}^{(s)} \leftarrow x_{i,j}^{*}$
      else
        $x_{i,j}^{(s)} \leftarrow x_{i,j}^{(s-1)}$
      end if
    end for
    for signal fraction parameters $k = 1$ to $k = N_{comp} - 1$ **do**
      $\epsilon_{f_{i,k}} \leftarrow N(f_{i,k}^{(s-1)}, w_{f_{i,k}})$
      $f_{i,k}^{*} \leftarrow f_{i,k}^{(s-1)} + \epsilon_{f_{i,k}}$
      calculate $\alpha(f_{i,k} \rightarrow f_{i,k}^{*})$ from Equation (17)
      $u \leftarrow$ sample from unif$(0, 1)$
      if $u < \alpha$ **then**
        $f_{i,k}^{(s)} \leftarrow f_{i,k}^{*}$
      else
        $f_{i,k}^{(s)} \leftarrow f_{i,k}^{(s-1)}$
      end if
    end for
    $f_{i,N_{comp}}^{(s)} \leftarrow 1 - \sum_{k=1}^{N_{comp}-1} f_{i,k}$
    if $s \mod 100 = 0$ **then**
      for voxels $i = 1$ to $i = N$ **do**
        Update $\{w_{x_{i,j}}\}_{j=1}^{J}$ and $\{w_{f_{i,k}}\}_{k=1}^{N_{comp}-1}$ using Equation (18)
      end for
    end if
  end for
end for

Bayesian priors were defined over the whole phantom excluding the background (i.e. one ROI).

We then applied our Bayesian model fitting approach on publicly-available data provided by the HCP WU-Minn Consortium (48 Subjects Test

Retest Data Release, release date: Mar 01, 2017, available online at http://humanconnectome.org). Data from a single subject was used. The white matter (WM), cortical gray matter (GM), sub-cortical GM and ventricle ROIs derived from the Freesurfer T_1 segmentations (these provided the best contrast between tissues) were transformed into diffusion space via linear and non-linear registration between the subject's T_1-weighted image and the $b = 0$ dMRI data. The MCMC algorithm was applied with the Bayesian priors defined over these four ROIs.

3 Results

Figure 2 displays the output of several runs of the MCMC algorithm for a single voxel in the Shepp-Logan data. The MCMC chains and posterior distributions demonstrate that voxelwise parameter estimates converged to the ground truth value under a range of perturbations. Figure 3 compares the least squares and Bayesian parameter maps with the ground truth. The Bayesian approach more accurately replicated the ground truth and provided lower errors than least squares approach, particularly in low SNR cases.

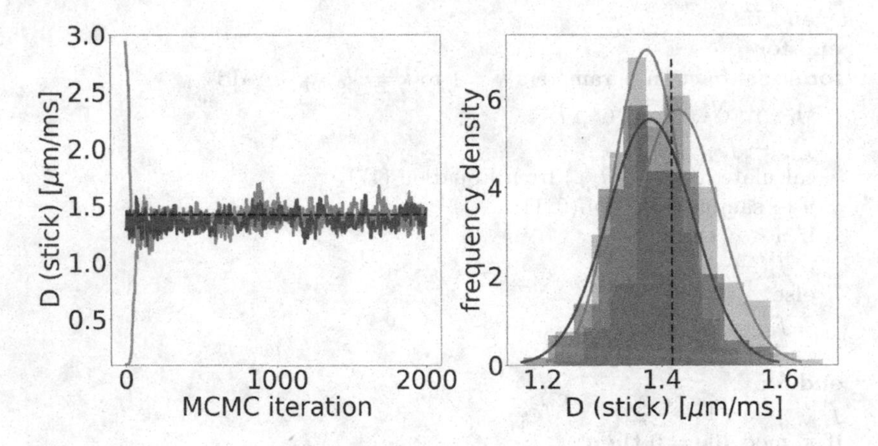

Fig. 2. MCMC output from the Bayesian ball-stick model fit on the Shepp-Logan phantom data. The left panel shows three MCMC chains for the stick parallel diffusivity in a single voxel; the initial parameter value in each run was given a different perturbation. The right panel shows the posterior distribution from each run, calculated on all samples after the burn-in of 1000 steps. The ground truth parameter value is indicated by the black lines. (Color figure online)

Figure 4 shows the MCMC algorithm results on the HCP data. The Bayesian fit clearly removed some apparent outlier voxels when compared to the least squares fit (see arrows).

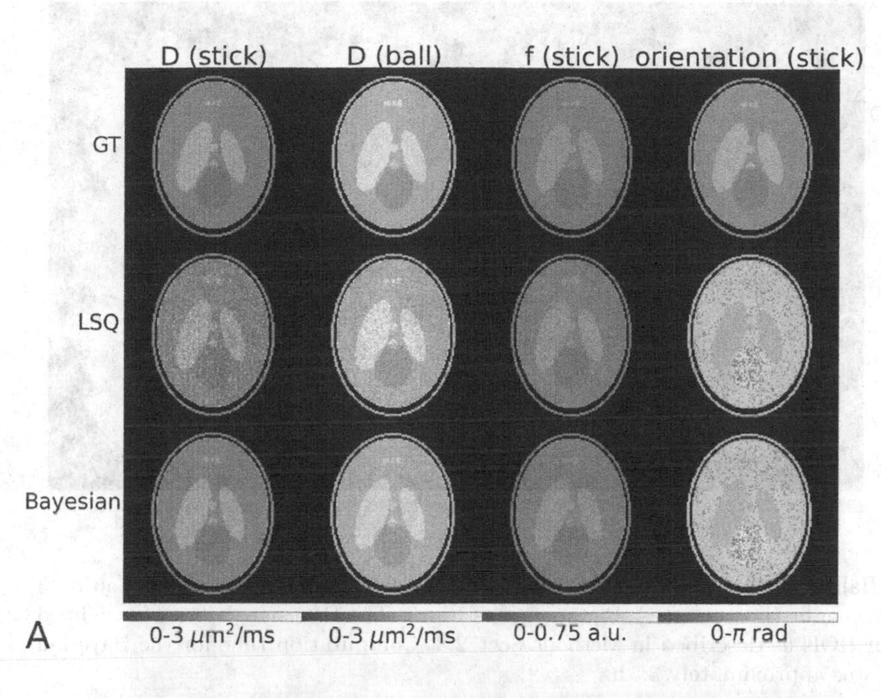

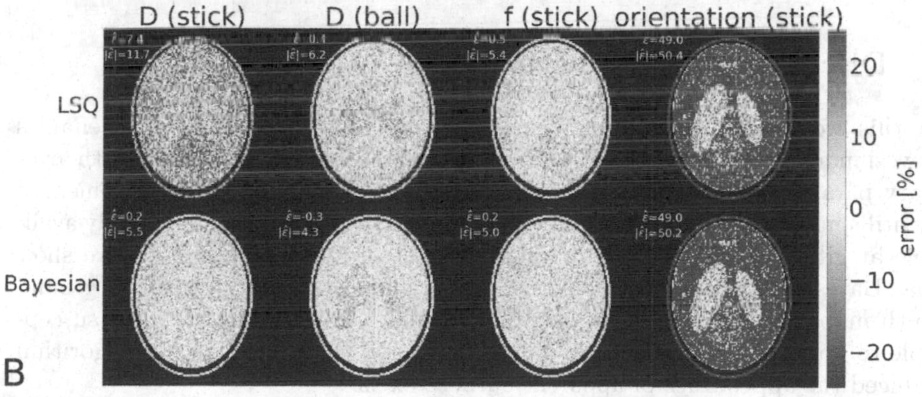

Fig. 3. Ball-stick model parameter maps in the Shepp-Logan phantom synthetic images. A. Ground truth (top row), least-squares derived (middle row) and Bayesian derived (bottom row) parameter maps for the stick parallel diffusivity (D_{par}) in μm^2/ms, ball isotropic diffusivity (D_{iso}) in μm^2/ms, stick signal fraction (f_{par}) and elevation orientation parameter (ϕ_1) in radians. B. Relative error maps for the least-squares fits (top row) and Bayesian fits (bottom row). Bayesian priors were defined over the whole image. The mean relative error ($\hat{\epsilon}$) and mean absolute relative error ($|\hat{\epsilon}|$) are displayed for each fitted parameter (both in %). Computation time for the Bayesian method was approximately 1 h.

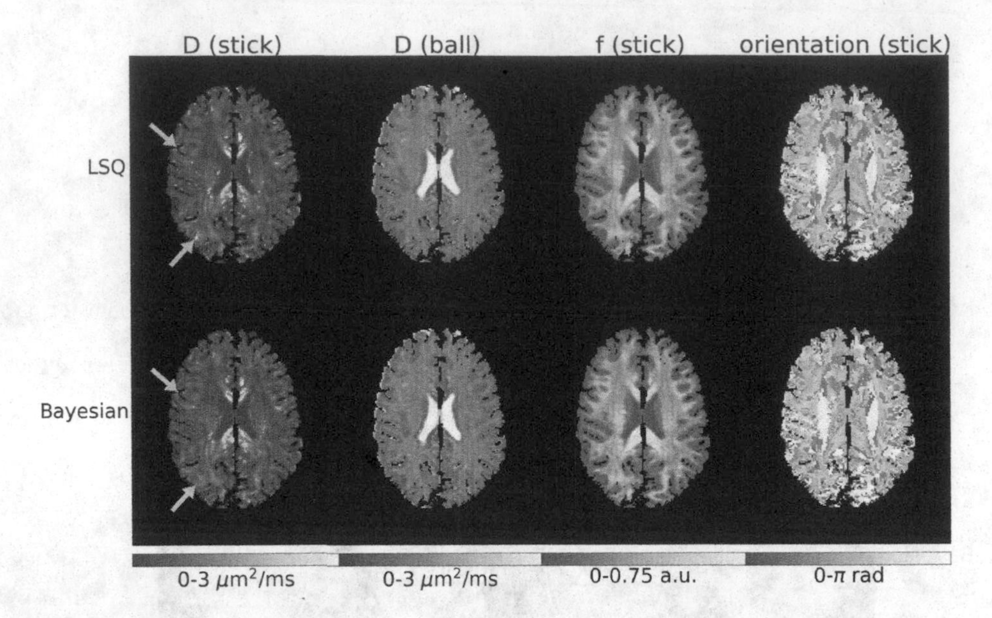

Fig. 4. Ball-stick parameter maps in the HCP data for the least squares fit (top row) and Bayesian fit (bottom row). Parameters as Fig. 3. The Bayesian priors were defined over four ROIs as described in Methods Sect. 2.8. Computation time for the Bayesian method was approximately 3.5 h.

4 Discussion and Conclusions

In this work we present an extension to previous approaches that enables Bayesian hierarchical model fitting for a general microstructural model with arbitrary parameter constraints and regional priors. The algorithm is implemented by utilising and adapting the Dmipy software package, and is made publicly available at github.com/PaddySlator/dmipy-bayesian. On synthetic data, we show that Bayesian fitting of the ball-stick model more accurately recovered ground truth maps than least squares fitting, particularly for parameters more susceptible to noise such as the stick parallel diffusivity. On HCP data, the algorithm reduced the appearance of apparent outlier voxels.

Although high SNR images from the HCP were used as the test data here, we anticipate the biggest gains of this approach will be seen in lower SNR data and more complex models. This may enable estimation of richer microstructural detail in a greatly reduced acquisition time.

The algorithm has limitations that motivate future work. The Bayesian approach assumes Gaussian noise, which may not be appropriate in all cases, particularly in low SNR cases. We also assumed that a Gaussian prior is suitable for all parameters; however, this may not be the optimal choice, particularly for orientation parameters where a flat prior may be more appropriate and could enable improved orientation parameter maps over LSQ (our current Bayesian implementation doesn't dramatically improve orientation maps, see Figs. 3 and

4). Alternative prior choices are an avenue for future work. Utilising probabilistic segmentations, rather than the current hard-thresholded ROIs which may bias parameter estimation in partial volume voxels, is possible, but may complicate the MCMC inference. More complex model choices, as well as comparisons with alternative model fitting methods (e.g. Harms et al. [3]), should also be explored to better quantify the benefits of our Bayesian approach.

To conclude, we derive a general Bayesian hierarchical microstructural model and an MCMC algorithm for model inference given dMRI data. The algorithm, and corresponding open-source software, newly enables Bayesian model fitting for a wide range of microstructure imaging techniques.

Acknowledgements. This work was supported by EPSRC grants EP/M020533/1, EP/V034537/1 and EP/S031510/1, as well as the NIHR UCLH Biomedical Research Centre. The authors would like to thank the organisers of the Brainhack Micro2Macro 2021 hackathon, and the International Global Brainhack 2020.

References

1. Fick, R.H., Wassermann, D., Deriche, R.: The dmipy toolbox: diffusion MRI multi-compartment modeling and microstructure recovery made easy. Front. Neuroinf. **13**(October), 1–26 (2019)
2. Flouri, D., et al.: Improved fetal blood oxygenation and placental estimated measurements of diffusion-weighted MRI using data-driven Bayesian modeling. Magn. Reson. Med. **83**(6), 2160–2172 (2020)
3. Harms, R.L., Roebroeck, A.: Robust and fast markov chain monte carlo sampling of diffusion MRI microstructure models. Front. Neuroinf. p. 97 (dec)
4. Kaden, E., Kelm, N.D., Carson, R.P., Does, M.D., Alexander, D.C.: Multi-compartment microscopic diffusion imaging. NeuroImage **139**, 346–359 (2016)
5. Koh, D.M., Collins, D.J., Orton, M.R.: Intravoxel incoherent motion in body diffusion-weighted MRI: reality and challenges. Am. J. Roentgenol. **6**, 1351–1361 (2011)
6. Le Bihan, D., Breton, E., Lallemand, D., Aubin, M.L.L., Vignaud, J., Laval-Jeantet, M.: Separation of diffusion and perfusion in intravoxel incoherent motion MR imaging. Radiology **2**, 497–505 (1988)
7. Novikov, D.S., Fieremans, E., Jespersen, S.N., Kiselev, V.G.: Quantifying brain microstructure with diffusion MRI: Theory and parameter estimation. NMR Biomed. **4**, e3998 (2019)
8. Orton, M.R., Collins, D.J., Koh, D.M.M., Leach, M.O.: Improved intravoxel incoherent motion analysis of diffusion weighted imaging by data driven Bayesian modeling. Magn. Reson. Med. **1**, 411–420 (2014)
9. Panagiotaki, E., et al.: Microstructural characterization of normal and malignant human prostate tissue with vascular, extracellular, and restricted diffusion for cytometry in tumours magnetic resonance imaging. Invest. Radiol. **50**(4), 218–227 (2015)
10. Zhang, H., Schneider, T., Wheeler-Kingshott, C.A., Alexander, D.C.: NODDI: Practical in vivo neurite orientation dispersion and density imaging of the human brain. NeuroImage **4**, 1000–1016 (2012)

Brain Tissue Microstructure Characterization Using dMRI Based Autoencoder Neural-Networks

Mauro Zucchelli(✉), Samuel Deslauriers-Gauthier, and Rachid Deriche

Inria, Université Côte d'Azur, Sophia-Antipolis, France
{mauro.zucchelli,samuel.deslauriers-gauthier,rachid.deriche}@inria.fr

Abstract. In recent years, multi-compartmental models have been widely used to try to characterize brain tissue microstructure from Diffusion Magnetic Resonance Imaging (dMRI) data. One of the main drawbacks of this approach is that the number of microstructural features needs to be decided a priori and it is embedded in the model definition. However, the number of microstructural features which is possible to obtain from dMRI data given the acquisition scheme is still not clear.

In this work, we aim at characterizing brain tissue using autoencoder neural networks in combination with rotation-invariant features. By changing the number of neurons in the autoencoder latent-space, we can effectively control the number of microstructural features that we obtained from the data. By plotting the autoencoder reconstruction error to the number of features we were able to find the optimal trade-off between data fidelity and the number of microstructural features. Our results show how this number is impacted by the number of shells and the *b*-values used to sample the dMRI signal. We also show how our technique paves the way to a richer characterization of the brain tissue microstructure in-vivo.

Keywords: Spherical harmonics · Rotation invariant features · Neural networks · Diffusion MRI · Autoencoder

1 Introduction

Diffusion Magnetic Resonance Imaging (dMRI) has been widely used in research to probe the brain tissue microstructure in the living human brain. To estimate microstructural features from the dMRI signal, it is first necessary to select a biophysical model which can capture the desired features [7]. The choice of the model is limited by the type of dMRI data available, e.g. the type of diffusion sequence, the number of gradients (or shells), the number of echo times, etc. [5]. Although this approach has been successful, it has several limitations. The biggest limitation is that the choice of the model completely decides the number and type of microstructural feature which are obtainable from the data [8]. A model-fitting will always provide a result but the accuracy of this result is not

S. Cetin-Karayumak et al. (Eds.): CDMRI 2021, LNCS 13006, pp. 48–57, 2021.
https://doi.org/10.1007/978-3-030-87615-9_5

simple to verify. Nowadays, several software tools [1,2] easily permit researchers to fit multiple models on the same data. However, the comparison and validation of such models is still an open question[1].

Neural networks (NN) have become widely popular in medical imaging for their astonishing classification given a sufficiently large and accurate training set [6]. One of the most important features of NN is that they can be considered a model-free approach. Normally, no information on the characteristics of the data that we want to classify is given to the NN, because they are able to learn the classes representation directly from the training data.

NN have been already applied to dMRI data in order to obtain microstructural features [3,15]. In these works, a multi-compartment model has been chosen for generating the microstructural features used as labels for the NN training set. This approach limits the representation power of the NN to the model used for labeling the training data, not fully exploiting the model-free characteristic of the NN.

Autoencoders are a particular family of NN, principally used for denoising and feature extraction [12]. An autoencoder NN is composed of two parts, an encoder, and a decoder. The encoder that takes as input the data and outputs a certain number of features (also known as latent-space), corresponding to the number of neurons in the last layer of the encoder. The decoder takes as input these features and outputs the original data (see Fig. 1). An autoencoder NN is trained by trying to minimize the error between the data and the output of the decoder. In order to prevent the network from learning the identity operation, the number of features produced by the encoder is strictly lower than the size of the input data. This enables the autoencoder NN to codify each input in a small number of features which can then be used for classification or other tasks. Autoencoders have been recently used in dMRI by Vasilev and colleagues to detect abnormal voxels in multiple-sclerosis patients' data [12]. One of the main limitations of using autoencoders with dMRI data is that the dMRI signal is strongly directional-dependent and even a slight change in the orientation in the underlying brain fiber bundles will result in a completely different signal profile. This means that if the dMRI signal samples are used for training the autoencoder at least some of its features will be used to capture the orientation profile. Tissue orientation can be easily retrieved from dMRI data using the popular spherical deconvolution approaches [11], therefore it is generally not included in the scope of microstructural estimation technique. In fact, Rotation Invariant Features (RIF) can be used as an input microstructural models, instead of the row signal samples [9,14].

In this article, we aim to combine high order RIF, derived from the 4th order Spherical Harmonics expansion of the dMRI signal, with autoencoders. Our purpose is to characterize the number of microstructural features which is possible to obtain from dMRI pulse-gradient spin-echo (PGSE) data, regardless of the tissue orientation. We also investigate how this number changes to the number of shells used to sample the data. Our results show that this technique

[1] see the recent MEMENTO challenge: https://my.vanderbilt.edu/memento/.

is indeed feasible to model dMRI data and show also that increasing the number of shells and the b-value provide a richer characterization of the brain tissue microstructure in-vivo.

2 Materials and Methods

2.1 Rotation-Invariant Features of the dMRI Signal

The dMRI signal at each b-value can be represented as a spherical function $f(\mathbf{u})$, where the unit vector $\mathbf{u}$ corresponds to the gradient direction. Considering the diffusion signal is real-valued and antipodal-symmetric, we can expand it into a truncated Spherical Harmonics (SH) series

$$f(\mathbf{u}) = \sum_{l=0,even}^{L} \sum_{m=-l}^{l} c_{lm} Y_l^m(\mathbf{u}) \tag{1}$$

where c_{lm} are the real-valued SH coefficients Y_l^m are the real SH [4] of degree l and order m, and L is the maximum degree. In [14], the authors defined a set of algebraic independent RIF. Each invariant can be calculated as

$$I_{\mathbf{l}}[f] = \sum_{m_1=-l_1}^{l_1} \cdots \sum_{m_d=-l_d}^{l_d} c_{l_1 m_1} \cdots c_{l_d m_d} G(l_1, m_1 | \cdots | l_d, m_d) \tag{2}$$

where G are the generalized Gaunt coefficients [4], namely the integral of d SH and the vector $\mathbf{l}$ is equal to $[l_1, l_2 \cdots, l_d]$. Equation (2) provides a rotation invariant indices for any set of $\mathbf{l}$. In this work, we focused on the 12 algebraic independent RIF obtained from the SH expansion at $L = 4$ of the diffusion signal [14]. Being invariants, these RIF depends only on the "shape" of the signal and not on the orientation. For example, the value of the RIF change between crossing fibers and single fiber voxels, if two voxels present different intra-axonal signal fractions, or if the water diffusivity changes between the voxels [9]. Since multiple factors contribute to the invariants numerical values they cannot be considered biomarkers per se, but by removing the orientation-dependency they greatly improve the estimation of brain tissue microstructure. In particular, neural network-based microstructural feature estimation considerably benefits from this form of rotation invariant signal representation [15].

2.2 In-Vivo Dataset

In this work, we consider three subjects of Human Connectome Project (HCP) diffusion MRI young-adults dataset [10]. For each subject, 288 dMRI volumes has been acquired: 18 volumes at b-value $0\,\mathrm{s/mm^2}$, 90 volumes at b-values $1000\,\mathrm{s/mm^2}$, 90 volumes at b-values $2000\,\mathrm{s/mm^2}$, and 90 volumes at b-values $3000\,\mathrm{s/mm^2}$. Each volume is composed of $145 \times 174 \times 145$ voxels with a resolution of $1.25 \times 1.25 \times 1.25\,\mathrm{mm^3}$. Before calculating the SH expansion, we normalized each volume dividing the diffusion signal by the mean of the b-value $0\,\mathrm{s/mm^2}$ volumes.

2.3 Autoencoder Design

Figure 1 shows a graphical representation of the autoencoder used in this work. We design our autoencoder as a fully connected NN for taking as input the 12 RIF calculated from each shell of the HCP dataset. Therefore, the number of input (and output) channels corresponds to 12 for the single-shell b = $1000\,\mathrm{s/mm^2}$ data, 24 for the two shells up to b = $2000\,\mathrm{s/mm^2}$, and 36 for the complete dataset with three shells up to b = $3000\,\mathrm{s/mm^2}$. Following the input layer, we add a first hidden layer composed of 50 neurons combined with a Rectified Linear Unit (ReLU) functions. After this layer, the data is passed to another layer of neurons plus ReLU forming the latent-space layer. The output of the latent-space layer represents the set of microstructural features that characterize the dMRI signal. The cardinality of the latent-space layer is the main variable of interest for our autoencoder design. This number defines the amount of features that the autoencoder uses to represent the input. In this work, we created several neural networks with latent-space ranging from 3 neurons up to the number of the input channels. In this last case, the network should be able to simulate the identity transform perfectly recovering the input, therefore we use it only as a reference limit-case. These first three elements form the *encoder*. The *decoder* takes as input features output of the latent space and passes it to the second hidden layer composed of 50 neurons. As in the case of the first hidden layer, each neuron is followed by a ReLU function. The output of the second hidden layer is forwarded to the neurons of the output layer having the same number as the input of the network. Each autoencoder is optimized trying to minimize the Mean Square Error (MSE) between the input RIF at the different shell and the output of the network. Since each RIF have a different range of values, we scale each invariant according to a factor of 1, 10, or 100 to bring them in the same numerical range. This normalization is necessary in order to avoid that the biggest invariant (I_0, the mean of the signal) dominates the optimization reducing the influence of the other RIF. We trained the network using each of the three HCP subjects, independently, splitting the voxel randomly. Sixty percent of each brain voxels were used for training the network, and forty percent for testing. To test the robustness of the method, we used 10 different random seeds for dividing the data into training and testing, and for the network initialization and optimization. All the autoencoders implemented in this work are written in python using the Pytorch[2] library and trained using the Adam optimizer with a learning rate of 5×10^{-5} and a weight decay of 1×10^{-6}. The training set is split in batches of 1000 elements and we trained the networks for 300 epochs. All of the autoencoder hyper-parameters (including the number of layers, the learning rate, and the weight decay), were optimized by minimizing the MSE of the invariants on one of the HCP subjects.

[2] https://pytorch.org/.

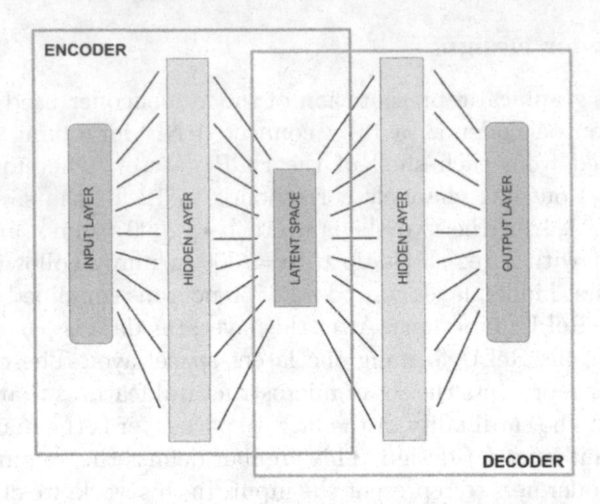

Fig. 1. Graphical representation of the autoencoder design used in this work. The input layer counts 12 × number of *b*-values channels, each hidden layer is composed of 50 neurons, and the output layer has the same size of the input. For each data we trained several autoencoder with the number of neurons in the latent space ranging from 3 to the size of the input channel.

3 Results and Discussion

In this work, we used the RIF extracted from the different shells of the dMRI signal as input, and the autoencoder latent-space represents the microstructural features. However, it is unclear how many microstructural features is it possible to extract from PGSE multi-shell dMRI data, in particular, as a function of the number of shells and the *b*-values used in the acquisitions. Figure 2 shows the log MSE between the original and reconstructed RIF for the testing set voxels of three HCP subjects (in red, green, and blue) as a function of the number of neurons in the latent-space. Each transparent line represents one random-seed instance of the same network, and the solid line the average of the ten networks results. Figure 2 (left) shows the results for the autoencoders trained using the 12 RIF at *b*-value $1000\,\mathrm{s/mm^2}$. At this b-value, the difference in the reconstruction error between the subjects is in the same range of the changes obtained training the same subject with a different network initialization, in particular, when the number of features of the latent-space is lower than 9. We calculated the elbow-point of the average of the curves for each subject, and we mark it on the graph with black crosses. The elbow-point represents an empirical way to find the "optimal" trade-off between reconstruction error and the number of features. In our experiments, two subjects (160123 and 473952) have reached an optimal number of features with 6 neurons in the latent-space, and one subject (174437) with 9 neurons. These features contain information on both the inner properties of the axons and extracellular space (like the axonal density and water diffusivity) and information regarding the axons architecture (crossing bundles, fibers fanning, etc.). This number of features is higher than the information usually retrieved from single-shell data, which are generally modeled as the three eigenvalues of the diffusion tensor. However, in this case, we used SH of order 4 for modeling the signal, while the diffusion tensor can be considered an order 2 representation. Therefore, the increase in the number of features can be explained by the increase in the order of the signal representation. Figure 2 (center) shows the same graph for the two shells data at b-value 1000 and 2000 s/mm^2. In this case, we have 24 RIF as input (12 for each shell). The autoencoders show a different reconstruction error for each subject when the latent-space grows bigger than 15 neurons. Before this point, the difference between the MSE of the different subjects is in the range of variance induced by the change of the random seed. The optimal number of features, calculated by finding the elbow-points of the curves, increase to 13 for subjects 160123 and 473952 and 14 for subject 174437. This means that there is a minimum increase of 5 features while moving from single-shell low-b-value data, to two shells data. Training the network with all the three shells, up to b-value 3000 s/mm^2 further increase this difference (see Fig. 2 right). In this case, we have 36 input channels considering 12 RIF per shell. Using the b-value 3000 s/mm^2 data we observe a clear difference between the reconstruction error of the three subjects. As in the previous cases, this difference increases with the increase of the number of features of the latent-space. It is necessary to underline that each subject has been used for training its autoencoder independently to the other two subjects. This means that there is something in the b-value 3000 s/mm^2 RIF which led the autoencoder to have a systematic different reconstruction error between the subject. This can be due to noise or another hidden acquisition artifact that may be completely uncorrelated to tissue microstructure. Although the errors are different, the trend of the errors is very close to each other. In fact, all three subjects have an optimal number of features at 15 latent-space neurons. The increase to the two-shells data is reduced to only one additional microstructural feature. This means that adding a third shell at b-value 3000 s/mm^2 while adding some subject-specificity and reduce the variance of the results does not increase the number of microstructural features obtainable from the data sensitively. Although we do not have the data, we can imagine that with a further increase in the number of shells we will probably reach a plateau in the number of microstructural features. This result is in line with other studies whose highlight that in order to be sensitive to the full brain tissue microstructure it is necessary to include other dMRI acquisition data, for example, Spherical Encoding [5], shells acquired using multiple diffusion-times [7], or multiple echo-time [13].

Figure 3 shows three examples of the quality of the autoencoder RIF reconstruction for the three shells data with 15 neurons in the latent space (HCP subject 160123). The three considered RIF are the signal mean invariant I_0, the degree 2 power spectrum invariant I_{22}, and one of the new invariant developed by using the original framework presented in [14]: invariant I_{224}. As it is possible to see these three RIF were recovered almost perfectly by the autoencoder.

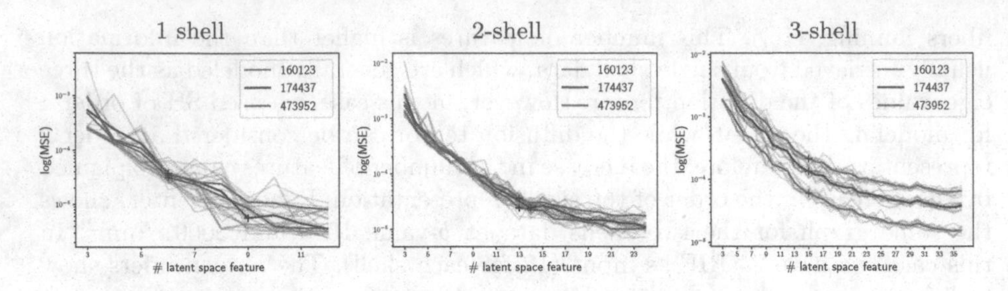

Fig. 2. Log MSE of the difference between the input RIF and the network RIF as a function of the size of the latent-space, for 1 shell (b-value = $1000\,\text{s/mm}^2$, on the left), two shells (up to b-value = $2000\,\text{s/mm}^2$, at the center), and three shells (up to b-value = $3000\,\text{s/mm}^2$, on the right). Each transparent line corresponds to a different random network initialization and the solid line represents the average for each of the three HCP subjects (in red, blue, and green). The black crosses correspond to the elbow point of the log curves. (Color figure online)

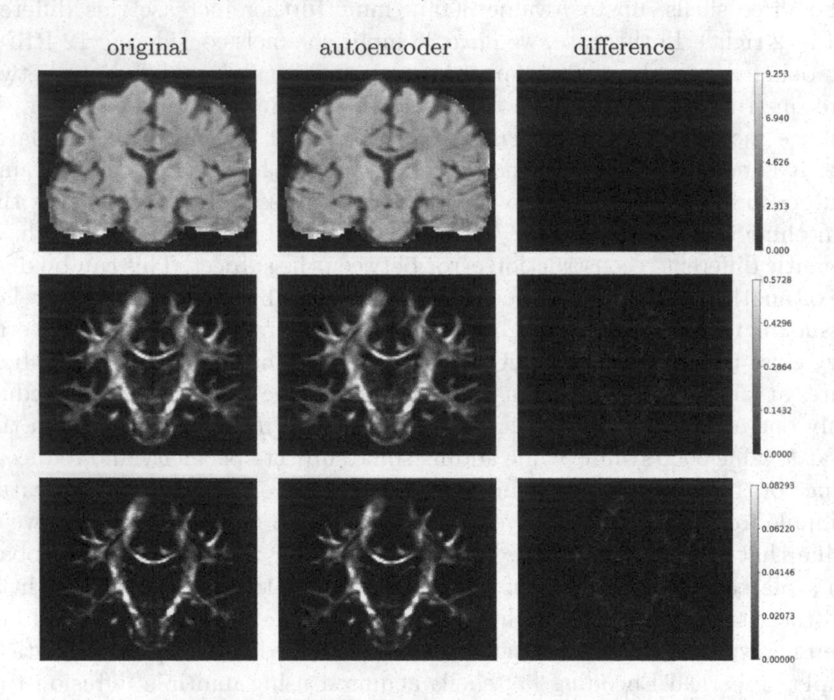

Fig. 3. Example of three RIF (left), their autoencoder reconstruction (center), and the absolute difference between the two (right). The top row represent invariant I_0 at b-value = $1000\,\text{s/mm}^2$, the second row invariant I_{22} at b-value = $2000\,\text{s/mm}^2$, and the third row invariant I_{224} at b-value = $3000\,\text{s/mm}^2$.

Although we only show three examples this level of quality is consistent for all the 36 RIF in the three shells.

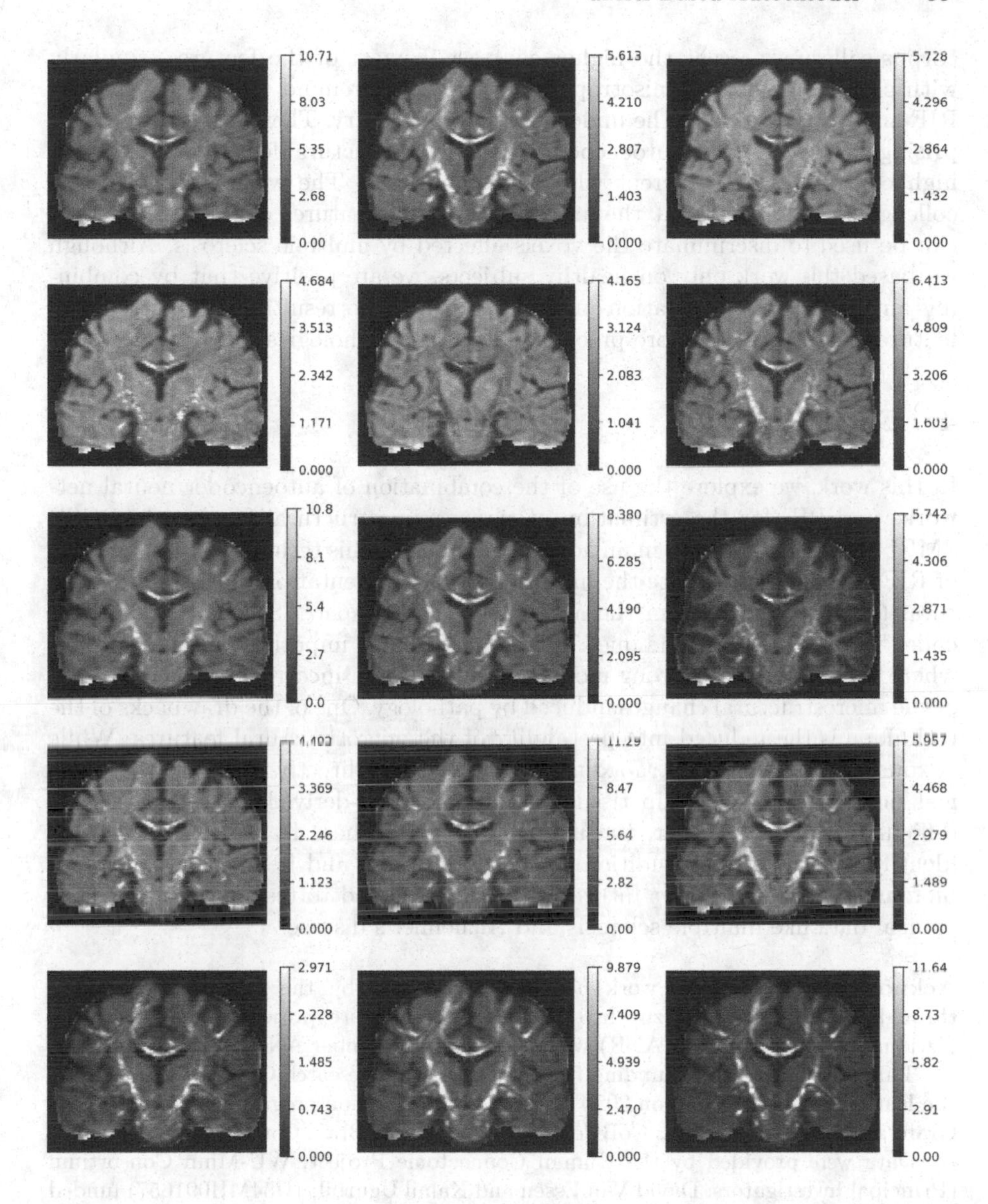

Fig. 4. The 15 autoencoder microstructural features obtained from the 3-shells data for HCP subject 160123.

Figure 4 shows the 15 microstructural features obtained for the HCP subject 160123 using the 3-shells RIF. As it is possible to see, the different features highlight different parts of the brain tissues. Most of these microstructural features present a high-contrast between white matter and gray matter. In particular, some single-fiber white-matter areas like the corticospinal tract and corpus-callosum present the higher contrast in most of the features, similarly with classical fractional anisotropy maps. We must remind the reader that the RIF are very sensitive to the underlying fiber geometry. Therefore, it is not surprising that most of the autoencoder-based microstructural features can provide high contrast in the different white-matter regions. The work of Vasilev and colleagues [12]] shows that the autoencoder-based features of the dMRI signal can be used to discriminate the voxels affected by multiple sclerosis. Although we based this work only on healthy subjects, we are positive that by combining autoencoders and rotation invariant features the resulting microstructural features might even be more precise in detecting pathological conditions.

4 Conclusions

In this work, we explore the use of the combination of autoencoder neural networks and RIF for the estimation of tissue microstructural features in PGSE dMRI data. There are two main advantages of using this technique. First, the use of RIF permits to eliminate the influence of tissue orientation in the microstructural features derived from the autoencoder latent-space. Second, the autoencoder is a model-free technique. This is particularly important in clinical data where the choice of the wrong model could lead to an incorrect characterization of the microstructural changes induced by pathology. One of the drawbacks of the technique is the reduced interpretability of the microstructural features. While "axonal density" and "extra-axonal diffusivity" are directly linked to the physical properties of the brain tissue, the autoencoder-derived features are more difficult to label. However, the main objective of these features is to help to identify a pathological condition in the brain tissue and be effectively used as biomarkers. Therefore, our future work will be aimed to test our technique on clinical data like multiple sclerosis and Alzheimer's disease.

Acknowledgments. This work has been supported by the French government, through the 3IA Côte d'Azur Investments in the Future project managed by the National Research Agency (ANR) with the reference number ANR-19-P3IA-0002.

This work has received funding from the European Research Council (ERC) under the European Union's Horizon 2020 research and innovation program (ERC Advanced Grant agreement No 694665 : CoBCoM - Computational Brain Connectivity Mapping).

Data were provided by the Human Connectome Project, WU-Minn Consortium (Principal Investigators: David Van Essen and Kamil Ugurbil; 1U54MH091657) funded by the 16 NIH Institutes and Centers that support the NIH Blueprint for Neuroscience Research; and by the McDonnell Center for Systems Neuroscience at Washington University.

The authors are grateful to the OPAL infrastructure from Université Côte d'Azur for providing resources and support.

References

1. Fick, R.H., Wassermann, D., Deriche, R.: The Dmipy toolbox: diffusion MRI multi-compartment modeling and microstructure recovery made easy. Front. Neuroinform. **13**, 64 (2019)
2. Garyfallidis, E., et al.: Dipy, a library for the analysis of diffusion MRI data. Front. Neuroinform. **8**, 8 (2014)
3. Golkov, V., et al.: q-Space deep learning: twelve-fold shorter and model-free diffusion MRI scans. IEEE Trans. Med. Imag. **35**(5), 1344–1351 (2016). https://doi.org/10.1109/TMI.2016.2551324
4. Homeier, H.H., Steinborn, E.: Some properties of the coupling coefficients of real spherical harmonics and their relation to gaunt coefficients. J. Mol. Struct.: THEOCHEM **368**, 31–37 (1996). https://doi.org/10.1016/S0166-1280(96)90531-X, Proceedings of the Second Electronic Computational Chemistry Conference
5. Lampinen, B., Szczepankiewicz, F., Mrtensson, J., van Westen, D., Sundgren, P.C., Nilsson, M.: Neurite density imaging versus imaging of microscopic anisotropy in diffusion MRI: a model comparison using spherical tensor encoding. NeuroImage **147**, 517–531 (2017)
6. Liu, X., et al.: A comparison of deep learning performance against health-care professionals in detecting diseases from medical imaging: a systematic review and meta-analysis. Lancet Digital Health **1**(6), e271–e297 (2019)
7. Novikov, D.S., Fieremans, E., Jespersen, S.N., Kiselev, V.G.: Quantifying brain microstructure with diffusion MRI: theory and parameter estimation. NMR Biomed. **32**(4), e3998 (2019)
8. Novikov, D.S., Kiselev, V.G., Jespersen, S.N.: On modeling. Magn. Reson. Med. **79**(6), 3172–3193 (2018)
9. Novikov, D.S., Veraart, J., Jelescu, I.O., Fieremans, E.: Rotationally-invariant mapping of scalar and orientational metrics of neuronal microstructure with diffusion MRI. NeuroImage **174**, 518–538 (2018). https://doi.org/10.1016/j.neuroimage.2018.03.006
10. Sotiropoulos, S.N., et al.: Advances in diffusion MRI acquisition and processing in the human connectome project. Neuroimage **80**, 125–143 (2013)
11. Tournier, J.D., Calamante, F., Connelly, A.: Robust determination of the fibre orientation distribution in diffusion MRI: non-negativity constrained super-resolved spherical deconvolution. NeuroImage **35**(4), 1459–1472 (2007)
12. Vasilev, A., et al.: q-Space novelty detection with variational autoencoders. In: Bonet-Carne, E., Hutter, J., Palombo, M., Pizzolato, M., Sepehrband, F., Zhang, F. (eds.) Computational Diffusion MRI, pp. 113–124. Springer, Cham (2020). https://doi.org/10.1007/978-3-030-52893-5_10
13. Veraart, J., Novikov, D.S., Fieremans, E.: TE dependent diffusion imaging (TEdDI) distinguishes between compartmental T2 relaxation times. NeuroImage **182**, 360–369 (2018)
14. Zucchelli, M., Deslauriers-Gauthier, S., Deriche, R.: A computational framework for generating rotation invariant features and its application in diffusion MRI. Med. Image Anal. **60**, 101597 (2020)
15. Zucchelli, M., Deslauriers-Gauthier, S., Deriche, R.: Investigating the effect of DMRI signal representation on fully-connected neural networks brain tissue microstructure estimation. In: 2021 IEEE 18th International Symposium on Biomedical Imaging (ISBI), pp. 725–728 (2021). https://doi.org/10.1109/ISBI48211.2021.9434046

Synthesizing VERDICT Maps from Standard DWI Data Using GANs

Eleni Chiou[1,2(✉)], Vanya Valindria[1,2], Francesco Giganti[3,4], Shonit Punwani[5], Iasonas Kokkinos[2], and Eleftheria Panagiotaki[1,2]

[1] Centre for Medical Image Computing, UCL, London, UK
eleni.chiou.17@ucl.ac.uk
[2] Department of Computer Science, UCL, London, UK
[3] Department of Radiology, UCLH NHS Foundation Trust, London, UK
[4] Division of Surgery & Interventional Science, UCL, London, UK
[5] Centre for Medical Imaging, Division of Medicine, UCL, London, UK

Abstract. VERDICT maps have shown promising results in clinical settings discriminating normal from malignant tissue and identifying specific Gleason grades non-invasively. However, the quantitative estimation of VERDICT maps requires a specific diffusion-weighed imaging (DWI) acquisition. In this study we investigate the feasibility of synthesizing VERDICT maps from standard DWI data from multi-parametric (mp)-MRI by employing conditional generative adversarial networks (GANs). We use data from 67 patients who underwent both standard DWI-MRI and VERDICT MRI and rely on correlation analysis and mean squared error to quantitatively evaluate the quality of the synthetic VERDICT maps. Quantitative results show that the mean values of tumour areas in the synthetic and the real VERDICT maps were strongly correlated while qualitative results indicate that our method can generate realistic VERDICT maps that could supplement mp-MRI assessment for better diagnosis.

Keywords: VERDICT maps · DWI-MRI · Prostate Cancer · Generative adversarial networks (GANs)

1 Introduction

Multi-parametric (mp)-MRI, consisting of T2-weighted imaging, diffusion-weighted imaging (DWI) and dynamic contrast enhanced (DCE) imaging, provides non-invasive assessment of the prostate improving the detection and characterization of prostate cancer. However, despite its merits, mp-MRI has some important limitations. In particular, it is characterized by low specificity, provides equivocal findings for around 30% of the patients and correlates moderately with Gleason grade [1].

Towards addressing these limitations, advanced, model-based imaging techniques focus on extracting quantitative metrics that characterize the underlying tissue microstructure in-vivo by modeling the DWI signal [4]. In particular,

S. Cetin-Karayumak et al. (Eds.): CDMRI 2021, LNCS 13006, pp. 58–67, 2021.
https://doi.org/10.1007/978-3-030-87615-9_6

VERDICT (Vascular, Extracellular and Restricted Diffusion for Cytometry in Tumours) MRI [19,20], which has been recently in clinical trial [14] to supplement the standard mp-MRI for prostate cancer diagnosis, is a model-based, DWI technique that captures the main microstructural properties of cancerous tissue. VERDICT MRI has shown promising results discriminating normal from malignant tissue [8,9,19,25] and identifying specific Gleason grades in-vivo [15,26].

VERDICT MRI combines an optimized DWI acquisition protocol [18] and a mathematical model to estimate microstructural features such as cell size, density, and vascular volume fraction, all of which change in malignancy (Fig. 1). The general model characterizes water diffusion in three primary compartments allowing the estimation of intracellular (fIC), extracellular-extravascular (fEEs) and vascular (fVASC) volume fractions, and cell radius (R). However, the quantitative estimation of these parameters requires a specific DWI acquisition, different to the one widely used for the standard mp-MRI. Specifically, VERDICT MRI requires multiple and higher b-values with different diffusion times (90, 500, 1500, 2000, 3000 s/mm^2) in different directions to derive accurate estimates of the microstructure parameters. Higher b-values improve tumour conspicuity and characterization but require longer scan times. Thus, methods that can estimate the VERDICT maps using standard DWI data from mp-MRI acquisitions would be beneficial for improving diagnostic accuracy without increasing scan time and patient discomfort.

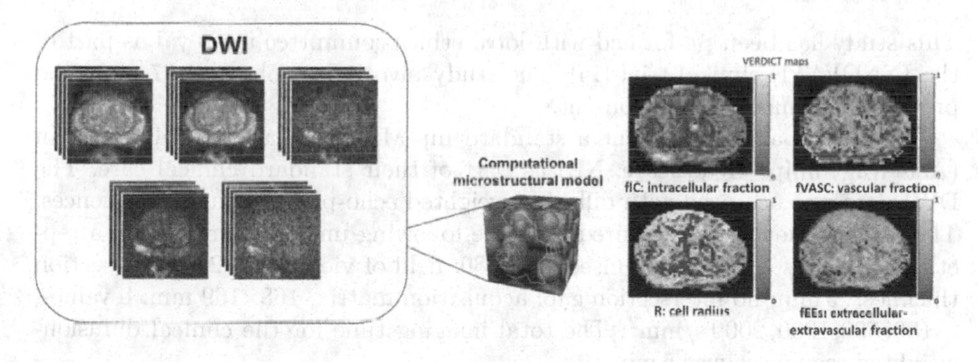

Fig. 1. VERDICT MRI framework. It combines an optimized diffusion-weighted imaging (DWI) protocol and a mathematical model to estimate microstructural features of tumours in-vivo.

Recently, several machine learning methods have been proposed to map an input image from one domain to an output image from a different domain aiming to improve the quality of data coming from routine, low-cost acquisitions or scanners or to eliminate the need for multi-modality scanning. Alexander et al. [2] proposed a general framework for image quality enhancement based on patch regression and demonstrated its effectiveness in super-resolution of brain diffusion tensor images and estimation of parametric maps from limited measurements. They further extended this approach with probabilistic deep learning formulation and showed that modeling uncertainty allows for better generalization [23,24]. Oktay et al. [17] proposed an image super-resolution approach based on a residual convolutional neural network (CNN) to reconstruct high resolution 3D volumes from a 2D image enabling more accurate analysis of cardiac morphology. In addition, approaches relying on generative adversarial networks (GANs) [11] have been proposed for super-resolution of structural brain MRI [7,22], endomicroscopy [21] and musculoskeletal MRI [6]. Nie et al. [16] proposed a GAN-based approach to generate CT images from MRI images to eliminate multi-modality scanning. Wolterink et al. [29] used CycleGAN [30] to translate MRI to CT in the absence of paired samples. Wang et al. [27] proposed a semi-supervised approach to synthesize ADC images from T2 images to boost performance of clinical tasks in settings where there is limited supervision. Chiou et al. [10] relied on stochastic translation to translate DWI from mp-MRI to raw diffusion VERDICT MRI to improve segmentation performance.

In this work we also rely on a GAN-based approach [12] to generate VERDICT maps from standard DWI data from clinical mp-MRI acquisitions to obtain microstructural information without requiring a specialized acquisition protocol.

2 Methods

2.1 Datasets

This study has been performed with local ethics committee approval as part of the INNOVATE clinical trial [14]. The study involved a cohort of 67 men who provided informed written consent.

All participants underwent a standard mp-MRI with a 3.0-T MRI system (Achieva, Philips Healthcare, NL) as part of their standard clinical care. The DWI data was acquired with diffusion-weighted echo-planar imaging sequences. The DWI sequence was acquired with the following imaging parameters: a repetition time msec/echo time msec, 2753/80; field of view, 220×220 mm; section thickness, 5 mm; no intersection gap; acquisition matrix, 168×169 mm; b values, $0, 150, 500, 1000, 2000\,\mathrm{s/mm^2}$. The total imaging time for the clinical diffusion-weighted sequences was 5 min 16 s.

VERDICT MRI data was acquired with pulsed-gradient spin-echo sequence (PGSE) using an optimised imaging protocol for VERDICT prostate characterization with 5 b-values ($90, 500, 1500, 2000, 3000\,\mathrm{s/mm^2}$) in 3 orthogonal directions [18]. Images with $b = 0\,\mathrm{s/mm^2}$ were also acquired for each b-value. The DWI sequence was acquired with the following imaging parameters: repetition time msec/echo time msec, 2482–3945/50–90; voxel size, $1.25 \times 1.25 \times 5\,\mathrm{mm^3}$; slice thickness, 5 mm; slices, 14; field of view, $220 \times 220\,\mathrm{mm^2}$. The images were reconstructed to a 176×176 matrix size. The total imaging time was 12 min 25 s.

VERDICT MRI maps were generated by using the accelerated microstructure imaging via convex optimization, or AMICO framework [3]. The model has three independent unknown parameters: fIC, R and fEES. fVASC is calculated as fVASC = 1 − fIC − fEES, and the diffusion and pseudodiffusion coefficients are fixed to dIC = dEES = $2 \times 10^{-9}\mathrm{m}^2/\mathrm{s}$, P = $8 \times 10^{-9}\mathrm{m}^2/\mathrm{s}$. As in [15], in this study we use the fIC, fEES, fVASC and R.

The regions of interest (ROIs) corresponding to Prostate Imaging Reporting and Data System (PI-RADS) [28] score 3, 4 and 5 were contoured on VERDICT MRI using mp-MRI for guidance by an experienced radiologist reporting more than 2000 prostate MR scans per year.

2.2 Proposed Model

Let $x \in \mathbb{R}^{H \times W \times C_{dwi}}$, where H and W are the height and the width of the DWI data and $C_{dwi} = 5$, the number of input channels corresponding the different b-values. Let also $y \in \mathcal{Y}^{H \times W \times C_{maps}}$, where $\mathcal{Y} = [0, 1]$, H and W are the height and the width of the VERDICT maps and $C_{maps} = 4$, the number of the maps. Our goal is to train a model which takes as input 2D DWI slices (5 b-values) and generates the corresponding VERDICT maps (4 maps).

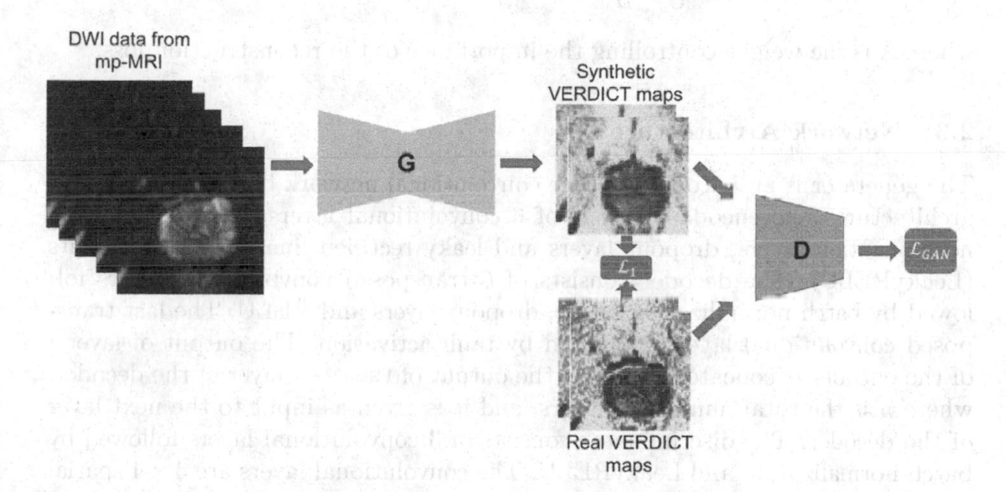

Fig. 2. Schematic representation of the proposed framework for synthesizing VERDICT maps form standard DWI data from mp-MRI acquisitions. The discriminator D is trained to discriminate between real and synthetic VERDICT maps while the generator G maps DWI data to synthetic VERDICT maps ($G : x \rightarrow y$) that cannot be distinguished from real VERDICT maps by the discriminator D. A GAN-type loss is used to push the distribution of the synthetic maps closer to the ground truth while the L1 loss ensures that the global and local structure of the synthetic maps do not deviate significantly from the real images.

In this work we use pix2pix framework [12], which has shown great success in natural images, to map standard DWI data from mp-MRI acquisitions to VERDICT maps. As it is illustrated in Fig. 2 the framework consists of a generator network G and a discriminator network D. The discriminator D is trained to discriminate between real and synthetic VERDICT maps while the generator G

maps DWI data to synthetic VERDICT maps ($G : x \rightarrow y$) that cannot be distinguished from real VERDICT maps by the discriminator D. The adversarial loss, L_{GAN}, can be expressed as

$$\mathcal{L}_{GAN}(G, D) = \mathbb{E}_{x,y}[\log(D(x, y))] + \mathbb{E}_{x}[\log(1 - D(G(x))]. \quad (1)$$

A generator trained solely using the adversarial loss function can synthesize realistic-looking maps which however do not preserve the global and local structure and deviate significantly from the real images. To address this problem and generate maps that both fool the discriminator and are close to the real ground truth maps we use a pixel reconstruction loss, i.e., $L1$ distance. The $\mathcal{L}_1$ loss can be expressed as

$$\mathcal{L}_1(G) = \mathbb{E}_{x,y}[\|G(x) - y\|_1]. \quad (2)$$

The final training objective can be written as

$$\min_{G} \max_{D} \mathcal{L}_{GAN}(G, D) + \lambda \mathcal{L}_1(G), \quad (3)$$

where λ is the weight controlling the importance of the reconstruction loss.

2.3 Network Architecture

The generator is an encoder-decoder convolutional network based on the U-Net architecture. The encoder consists of 6 convolutional layers followed by batch normalization layers, dropout layers and leaky rectified linear activation units (LeakyRELU). The decoder consists of 6 transposed convolutional layers followed by batch normalization layers, dropout layers and RELU. The last transposed convolutional layer is followed by tanh activation. The output of layer i of the encoder is concatenated with the output of the $n - i$ layer of the decoder, where n is the total number of layers, and it is given as input to the next layer of the decoder. The discriminator consists of 3 convolutional layers followed by batch normalization and LeakyRELU. The convolutional layers are 4×4 spatial filters applied with stride 2 and padding 1.

2.4 Implementation Details

We implement the framework using Pytorch. We train both the generator and discriminator networks using mini-batch stochastic gradient descent and apply the Adam solver with a mini-batch size of 32, and momentum parameters $\beta_1 = 0.5$, $\beta_2 = 0.999$. We train the networks for 10000 epochs with an initial learning rate at 0.0001 that starts decreasing linearly to 0 after 5000 epochs. We employ dropout as a regularization strategy with dropout rate 50%. We use 60% of the patients for training, 20% for validation and 20% for testing.

Table 1. Mean squared error (MSE) calculated on the entire maps and on the prostate region only for the four maps (fIC, fEES, fVASC, R). The results are given in mean (± std) format.

	MSE	MSE (prostate)
fIC	0.18 (0.05)	0.13 (0.03)
fEES	0.12 (0.04)	0.16 (0.04)
fVASC	0.21 (0.05)	0.18 (0.04)
R	0.23 (0.07)	0.19 (0.06)

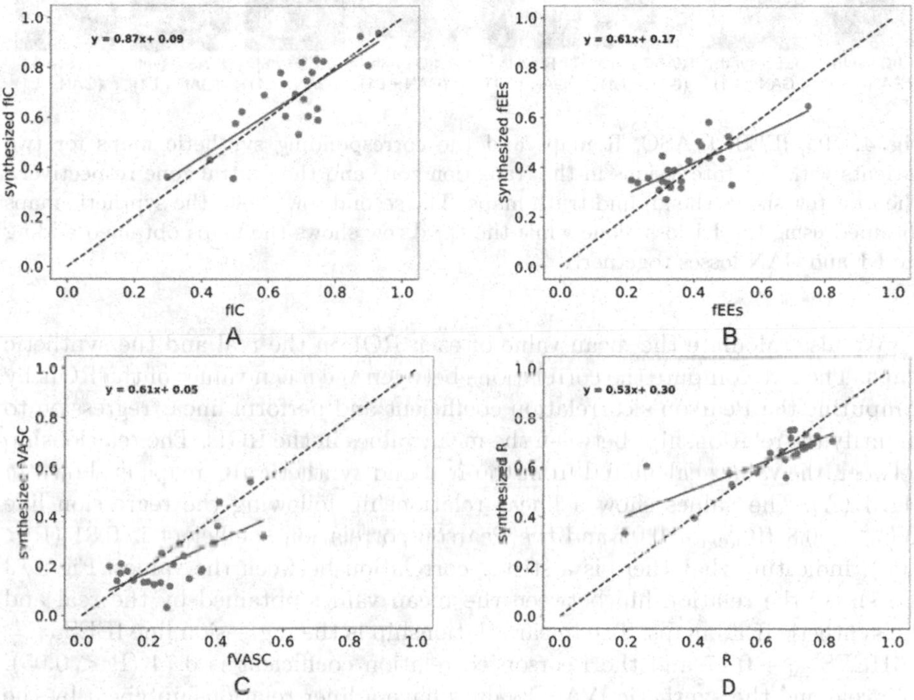

Fig. 3. A) Mean values of ROIs calculated from the real fIC as a function of the values calculated from the synthetic fIC. B) Mean values of ROIs calculated from the real fEES as a function of the values calculated from the synthetic fEES. C) Mean values of ROIs calculated from the real fVASC as a function of the values calculated from the synthetic fVASC. D) Mean values of ROIs calculated from the real R as a function of the values calculated from the synthetic R.

3 Results

We evaluate the quality of the synthetic maps based on the mean squared error (MSE). Table 1 shows the mean/std of the MSE over 13 test subjects computed on both the entire maps and the prostate region only.

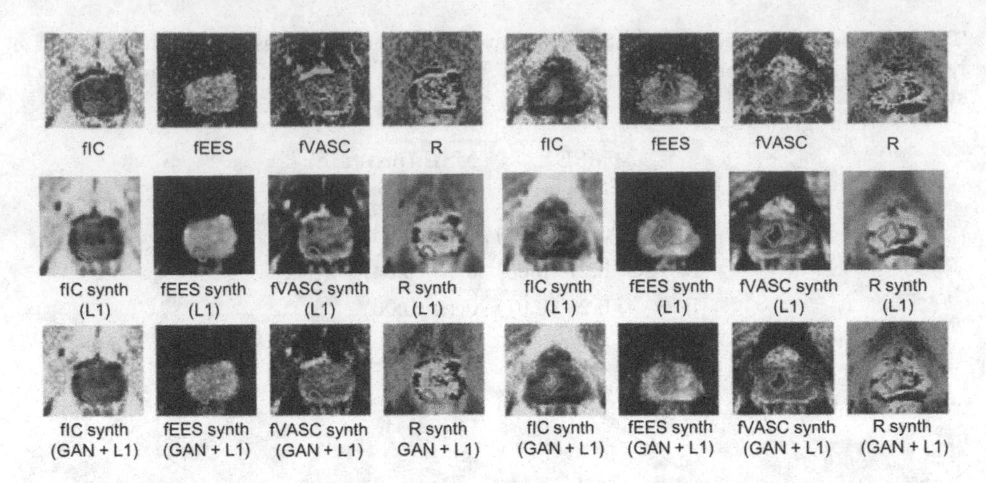

Fig. 4. fIC, fEES, fVASC, R maps and the corresponding synthetic maps for two patients with prostate lesions in the transition zone and the central zone respectively. The first row shows the ground truth maps. The second row shows the synthetic maps obtained using the L1 loss alone while the third row shows the maps obtained adding the L1 and GAN losses together.

We also calculate the mean value of each ROI on the real and the synthetic maps. Then we compute the correlations between the mean values of the ROIs by computing the Pearson's correlation coefficient and perform linear regression to quantify the relationships between the mean values in the ROIs. The relationship between the values calculated from the real and synthetic fIC maps is shown in Fig. 3 (A). The values show a linear relationship following the regression line $fIC_{syn} = 0.87fIC_{real} + 0.09$ and the Pearson correlation coefficient is 0.81 ($P < 0.05$), indicating that there is a strong correlation between the values. Figure 3 (B) shows the relationship between the mean values obtained by the real and the synthetic fEEs maps. The linear relationship is the regression line $fEES_{syn} = 0.61fEES_{real} + 0.17$ and the Pearson correlation coefficient is 0.74 ($P < 0.05$). The real and the synthetic fVASC values have a liner relationship given by the regression line $fVASC_{syn} = fVASC_{real}0.61 + 0.05$ and the Pearson correlation coefficient is 0.67 ($P < 0.05$). The real and the synthetic R values exhibit a linear relationship $R_{syn} = 0.53R_{real} + 0.30$ and the Pearson correlation coefficient is 0.82 ($P < 0.05$).

Figure 4 demonstrates an example of real and synthetic VERDICT parametric maps for two patients with prostate lesions in the transition zone and the central zone respectively. L1 loss alone leads to reasonable but blurry maps; adding the GAN loss gives much sharper results.

4 Discussion

In this study, we investigate a GAN-based approach for image synthesis for prostate cancer characterization. Our results demonstrate that the proposed approach is viable for generating realistic VERDICT maps from standard DWI data from mp-MRI acquisitions.

Table 1 gives the average MSE calculated between the synthetic and real VERDICT maps. As we can see in the table there is a small difference between the MSE computed using the whole image and using only the prostate region. This indicates that our approach is stable among all regions and all maps, especially the most important one (fIC) that has low error.

In addition the ROI measurements of the synthetic and real VERDICT maps are highly correlated. In particular, there is high correlation for fIC maps (0.81 ($P < 0.05$)), which have been shown to be the most important in differentiating specific Gleason grades.

Figure 4 shows that the synthetic maps have realistic appearance and preserve important quantitative information. In alignment with the real maps, the synthetic fIC, fEEs, fVASC and R maps clearly depict the lesions which are characterized by high signal in fIC and R maps and low signal in fEES and fVASC maps.

Our objective is to investigate the feasibility of generating VERDICT maps from the clinically-available DWI data from mp-MRI acquisitions to bring the advantages of VERDICT maps in clinical practice. We demonstrate that GAN-based methods have the potential to generate realistic VERDICT maps that preserve important clinical information without requiring specified DWI acquisition protocols, but only using the widely available DWI data from mp-MRI acquisitions. Obtaining VERDICT maps using standard DWI data would reduce acquisition time and patient discomfort. This would also allow us to use already acquired mp-MRI data to get microstructural information.

Despite the good quality of the synthetic maps there are still some limitations. Specifically, the synthetic images are smoother compared to real ones which means that for small ROIs the quantitative values could be wiped out. Methodological improvements that enforce semantic consistency before and after translation [5,13] could resolve this issue and allow the synthesis of high quality VERDICT maps.

5 Conclusion

In this work we present an approach for synthesizing realistic VERDICT maps from standard DWI data from prostate mp-MRI acquisitions. Our results indicate that the synthetic maps have realistic appearance and preserve important quantitative information. This could allow the exploitation of VERDICT maps for improved prostate cancer diagnosis without increasing acquisition time and patient discomfort.

References

1. Ahmed, H.U., et al.: Diagnostic accuracy of multi-parametric MRI and TRUS biopsy in prostate cancer (PROMIS): a paired validating confirmatory study. Lancet **389**, 815–822 (2017)

2. Alexander, D.C., et al.: Image quality transfer and applications in diffusion MRI. NeuroImage **152**, 283–298 (2017)
3. Bonet-Carne, E., et al.: VERDICT-AMICO: ultrafast fitting algorithm for non-invasive prostate microstructure characterization. NMR Biomed. **32**, e4019 (2019)
4. Bourne, R., Panagiotaki, E.: Limitations and prospects for diffusion-weighted MRI of the prostate. Diagnostics **6**, 21 (2016)
5. Cai, J., Zhang, Z., Cui, L., Zheng, Y., Yang, L.: Towards cross-modal organ translation and segmentation: a cycle and shape consistent generative adversarial network. MedIA **52**, 174–184 (2019)
6. Chaudhari, A.S., et al.: Super-resolution musculoskeletal MRI using deep learning. MRM **80**, 2139–2154 (2018)
7. Chen, Y., Shi, F., Christodoulou, A.G., Xie, Y., Zhou, Z., Li, D.: Efficient and accurate MRI super-resolution using a generative adversarial network and 3D multi-level densely connected network. In: Frangi, A., Schnabel, J., Davatzikos, C., Alberola-López, C., Fichtinger, G. (eds.) MICCAI 2018. LNCS, vol. 11070, pp. 91–99. Springer, Cham. https://doi.org/10.1007/978-3-030-00928-1_11
8. Chiou, E., Giganti, F., Bonet-Carne, E., Punwani, S., Kokkinos, I., Panagiotaki, E.: Prostate cancer classification on VERDICT DW-MRI using convolutional neural networks. In: MLMI (2018)
9. Chiou, E., Giganti, F., Punwani, S., Kokkinos, I., Panagiotaki, E.: Automatic classification of benign and malignant prostate lesions: a comparison using VERDICT DW-MRI and ADC maps. In: ISMRM (2019)
10. Chiou, E., Giganti, F., Punwani, S., Kokkinos, I., Panagiotaki, E.: Harnessing uncertainty in domain adaptation for MRI prostate lesion segmentation. In: Martel, A.L. et al. (eds.) MICCAI 2020. LNCS, vol. 12261, pp. 510–520. Springer, Cham (2020). https://doi.org/10.1007/978-3-030-59710-8_50
11. Goodfellow, I., et al.: Generative adversarial nets. In: NIPS (2014)
12. Isola, P., Zhu, J.Y., Zhou, T., Efros, A.A.: Image-to-image translation with conditional adversarial networks. In: CVPR (2017)
13. Jiang, J., et al.: Tumor-aware, adversarial domain adaptation from CT to MRI for lung cancer segmentation. In: Frangi, A., Schnabel, J., Davatzikos, C., Alberola-López, C., Fichtinger, G. (eds.) MICCAI 2018. LNCS, vol. 11071, pp. 777–785. Springer, Cham. https://doi.org/10.1007/978-3-030-00934-2_86
14. Johnston, E., et al.: INNOVATE: a prospective cohort study combining serum and urinary biomarkers with novel diffusion-weighted magnetic resonance imaging for the prediction and characterization of prostate cancer. BMC Cancer **16**, 1–11 (2016)
15. Johnston, E.W., et al.: VERDICT MRI for prostate cancer: intracellular volume fraction versus apparent diffusion coefficient. Radiology **291**, 391–397 (2019)
16. Nie, D., et al.: Medical image synthesis with deep convolutional adversarial networks. IEEE Trans. Biomed. Eng. **65**, 2720–2730 (2018)
17. Oktay, O., et al.: Multi-input cardiac image super-resolution using convolutional neural networks. In: Ourselin, S., Joskowicz, L., Sabuncu, M., Unal, G., Wells, W. (eds.) MICCAI 2016. LNCS, vol. 9902, pp. 246–254. Springer, Cham. https://doi.org/10.1007/978-3-319-46726-9_29
18. Panagiotaki, E., et al.: Optimised verdict MRI protocol for prostate cancer characterisation. In: ISMRM (2015)
19. Panagiotaki, E., et al.: Microstructural characterization of normal and malignant human prostate tissue with vascular, extracellular, and restricted diffusion for cytometry in tumours magnetic resonance imaging. Invest. Radiol. **50**, 218-227 (2015)

20. Panagiotaki, E., et al.: Noninvasive quantification of solid tumor microstructure using VERDICT MRI. Cancer Res. **74**, 1902–1912 (2014)
21. Ravì, D., Szczotka, A.B., Pereira, S.P., Vercauteren, T.: Adversarial training with cycle consistency for unsupervised super-resolution in endomicroscopy. MedIA **53**, 123–131 (2019)
22. Sánchez, I., Vilaplana Besler, V.: Brain MRI super-resolution using generative adversarial networks. In: MIDL (2018)
23. Tanno, R., et al.: Bayesian image quality transfer with CNNs: exploring uncertainty in dMRI super-resolution. In: Descoteaux, M., Maier-Hein, L., Franz, A., Jannin, P., Collins, D., Duchesne, S. (eds.) MICCAI 2017. LNCS, vol. 10433, pp. 611–619. Springer, Cham. https://doi.org/10.1007/978-3-319-66182-7_70
24. Tanno, R., et al.: Uncertainty modelling in deep learning for safer neuroimage enhancement: demonstration in diffusion MRI. NeuroImage **225**, 117366 (2021)
25. Valindria, V., Palombo, M., Chiou, E., Singh, S., Punwani, S., Panagiotaki, E.: Synthetic Q-space learning with deep regression networks for prostate cancer characterisation with verdict. In: ISBI (2021)
26. Valindria, V., et al.: Non-invasive Gleason score classification with VERDICT-MRI. In: ISMRM (2021)
27. Wang, Z., Lin, Y., Cheng, K.T.T., Yang, X.: Semi-supervised mp-MRI data synthesis with StitchLayer and auxiliary distance maximization. MedIA **59**, 101565 (2020)
28. Weinreb, J.C., et al.: PI-RADS Prostate Imaging Reporting and Data System: 2015, Version 2. European Urology (2016)
29. Wolterink, J.M., Dinkla, A.M., Savenije, M.H., Seevinck, P.R., van den Berg, C.A., Išgum, I.: Deep MR to CT synthesis using unpaired data. In: Tsaftaris, S., Gooya, A., Frangi, A., Prince, J. (eds.) SASHIMI 2017. LNCS, vol. 10557, pp. 14–23. Springer, Cham. https://doi.org/10.1007/978-3-319-68127-6_2
30. Zhu, J.Y., Park, T., Isola, P., Efros, A.A.: Unpaired image-to-image translation using cycle-consistent adversarial networks. In: ICCV (2017)

Tractography and Connectivity

A Novel Algorithm for Region-to-Region Tractography in Diffusion Tensor Imaging

Lars Smolders[1,2(✉)], Rick Sengers[1], Andrea Fuster[1], Mark de Berg[1], and Luc Florack[1]

[1] Department of Mathematics & Computer Science, Eindhoven University of Technology, 5600 MB, Eindhoven, The Netherlands
{H.J.C.E.Sengers,A.Fuster,M.T.d.Berg,L.M.J.Florack}@tue.nl

[2] Department of Neurosurgery, Elisabeth-Tweesteden Hospital, NL-5022 GC, Tilburg, The Netherlands
L.Smolders@etz.nl

Abstract. Geodesic tractography is an elegant, though typically time consuming method for finding connections or 'tracks' between given endpoints from diffusion-weighted MRI images, which can be representative of brain white matter fibers. In this work we consider the problem of constructing bundles of tracks between seed and target regions in the most efficient way. In contrast to streamline based methods, a naive region-to-region geodesic approach for finding the true bundle requires connecting all pairs of voxels in seed and target regions and then selecting the appropriate tracks. The running time of this approach is quadratic in the number of voxels, which is prohibitively long for clinical use. Moreover, matching full seed and target regions may include voxels that are not part of the target bundle, e.g. due to segmentation inaccuracies. In order to bring geodesic tractography closer to clinical applicability, we present a novel, efficient algorithm for region-to-region geodesic tractography which extends existing point-to-point algorithms and incorporates anatomical knowledge by assuming a topographic organization of fibers. The proposed method connects only seed and target voxels that belong to the target bundle, based on iterative refinement of a Delaunay tessellation of sample points. In addition, it can be used in combination with any point-to-point tractography algorithm. A theoretical analysis shows that, under reasonable assumptions, our algorithm is significantly more efficient than the quadratic-time solution. This is also confirmed by the experiments, which reveal a reduction in computation time of up to three orders of magnitude.

Keywords: Diffusion MRI · Geodesic tractography · Computational geometry

1 Introduction

In *tractography* one attempts to locate white matter bundles in-vivo without invasive surgery. Based on water diffusion profiles measured using diffusion-weighted magnetic resonance imaging (DWI), tentative fiber pathways producing

S. Cetin-Karayumak et al. (Eds.): CDMRI 2021, LNCS 13006, pp. 71–81, 2021.
https://doi.org/10.1007/978-3-030-87615-9_7

the observed diffusion profiles are reconstructed [5,15,17]. Recently a point-to-point geodesic tractography algorithm [10,19] based on Diffusion Tensor Imaging (DTI) was developed for obtaining individual fibers between two fiducial end-points. However, if one wishes to find the entire bundle of fibers connecting two three-dimensional brain regions, connecting all pairs of voxels in both regions is not a viable strategy. Firstly, calculating a track in geodesic tractography is computationally expensive, and becomes prohibitively costly when all possible connections between seed and target regions must be calculated before optimally connecting ones can be singled out, a complication due to the second order nature of the geodesic method ('geodesic completeness'). This problem will be worsened if DWI data resolution increases, further obstructing practical usage due to the quadratic increase in the number of connections to be computed. Furthermore, such a strategy might not be accurate enough, since it may produce fibers that are not part of the bundle, e.g. due to the seed and target regions not being specified with perfect accuracy or due to a point in one region not having any biologically plausible connections with points in the other region.

We present a novel efficient algorithm for region-to-region geodesic tractography alleviating both these issues. The goal is to find a collection of tracks connecting two regions in a way that optimizes a given measure of biological plausibility, while at the same time being able to distinguish parts of those regions that are likely misspecified, implying the need to score and rank (subsets of) tracks. A scoring criterion may incorporate prior knowledge based on anatomical expertise, or, as is the case here, be provided by the tractography algorithm. Computational reduction is achieved by an iterative coarse-to-fine sampling scheme that avoids the need to compute all possible tracks.

2 Theory and Algorithm

We depart from a DTI (matrix-valued) image D, a seed and target region S and T, both consisting of n square voxels[1] of unit size. We use the point-to-point tractography algorithm from [10] for two fiducial points in the brain to find a parameterized track $\gamma=\gamma(t)$ which minimizes the length functional $\mathcal{L}_g$ given by

$$\mathcal{L}_g(\gamma) = \int_0^1 \sqrt{\dot{\gamma}(t)^\intercal g(\gamma(t))\dot{\gamma}(t)}\, dt\,, \tag{1}$$

where $g \propto D^{-1}$ [11,15,17]. In this paper we choose $g = \mathrm{Adj}(D) = \det(D)D^{-1}$. In [3,10,18] a *connectivity score* is defined

$$C(\gamma) = \frac{\mathcal{L}_E(\gamma)}{\mathcal{L}_g(\gamma)}\,, \tag{2}$$

[1] For simplicity we present the algorithm for two sets of equal size n, but this can be replaced by sets of different sizes n_1 and n_2 by minor adjustments.

with $\mathcal{L}_E(\gamma)$ denoting the Euclidean length of the curve. This connectivity score captures the average apparent diffusivity along the track, so that a large score indicates that a path is globally well aligned with the main eigendirections of the diffusion tensors. The connectivity score (2) will be used throughout to assess tracks for their biological plausibility.

2.1 Iterative Refinement Algorithm

The aim of the *iterative refinement algorithm* is twofold. Firstly, the algorithm aims to identify regions $S_t \subset S$ and $T_t \subset T$ that are well connected to each other according to the connectivity score (2), and to construct the bundle of tracks connecting them; tracks that start or end in $S \setminus S_t$ and $T \setminus T_t$ will be discarded. Secondly, a speedup over simple brute-force computation of all n^2 connections should be realized. The key assumption allowing this speedup is that of *topographic organization*, which states that neighbouring points in one region connect with neighbouring points in the other region. While research is still ongoing in this area and the property does not hold uniformly for all functional brain regions [13], this assumption is reasonable within the context of single fiber bundles, e.g. the Optic Radiation (OR) bundle has been shown to have this property, where it is called 'retinotopy' or 'retinotopic mapping' [9,20,22], as well as the auditory system, where it is called 'tonotopy' [6,21]. Furthermore, algorithms based on this assumption have been successfully developed in the past, e.g. by Aydogan and Shi [4]. Since topographic organization implies that the boundary of the seed region is connected to the boundary of the target region, we set the goal of the algorithm to construct the set of tracks that start and end in the boundaries ∂S_t and ∂T_t, respectively, yielding a useful dimensionality reduction.

Additionally, in order to avoid computing all n^2 connections while searching for ∂S_t and ∂T_t, the algorithm operates on a coarser subset of sample points which is gradually refined. This introduces the notion of a *sampling scale* δ of a set of sample points P contained in a seed region S, defined as

$$\delta = \max_{s \in S} \min_{p \in P} \|s - p\| . \tag{3}$$

A sampling scale δ implies that a ball of radius δ centered around any point in the seed or target region must contain at least one sample point. We ensure a 'regular' spacing between sample points by applying the farthest-first traversal strategy, also used in Gonzalez's algorithm for the k-center problem [12]. Thus, starting from an initial sample point p_0, we generate sample points one by one, where the sample point p_i is chosen according to

$$p_i = \arg\max_{s \in S} \min_{0 \leq j < i} \|s - p_j\| . \tag{4}$$

Clearly, the distance that is maximized is precisely the current sampling scale δ, cf. Eq. (3). With each newly added sample point we either keep δ equal or reduce δ. The process of adding sample points terminates when we reach a desired sampling scale. Gonzalez's algorithm gives a constant-factor approximation of the minimal number of sample points needed to achieve a given δ [12]. For the target region Eqs. (3,4) are defined analogously with the replacement $S \rightarrow T$.

The iterative refinement algorithm employs a decreasing sequence of sampling scales $(\delta_i)_{i=0}^{k}$ to sample the regions S and T. At scale δ_0 we sample m points according to Eq. (4) in S and T, say $S_0 = (s_i)_{i=0}^{m}$ and $T_0 = (t_j)_{j=0}^{m}$. All points in S_0 are connected to all points in T_0 by geodesics $(\gamma_{i,j})_{i,j=1}^{m}$ with associated connectivity scores, cf. Eqs. (1, 2).

A point s_i is labeled '+' if

$$\max_{j=1,\ldots,m} C(\gamma_{i,j}) \geq \lambda C \doteq \lambda \max_{i,j=1,\ldots,m} C(\gamma_{i,j}) \tag{5}$$

and '−' otherwise, where $\lambda \in [0,1]$ is a relative threshold on C, the maximal connectivity found among all tracks. Points t_j are labeled similarly.

Two separate Delaunay tessellations [7] of S_0 and T_0 are constructed[2]. The critical observation is that, by virtue of topographic organization, the boundaries ∂S_t and ∂T_t must pass through tetrahedra that have both '+' and '−' vertices. Accordingly, we mark each tetrahedron as *active* if it has both '+' and '−' vertices, and *inactive* if all its vertices are either '+' or '−'. Next, we resample in all active tetrahedra using the finer sampling scale δ_1 and repeat this procedure until the finest scale δ_k has been used, at which point the optimal matching in terms of connectivity between the set of sample points sampled with scale δ_k is returned. To find this matching, we use the Munkres algorithm, also known as the Hungarian algorithm [14,16].

The running time of the iterative refinement algorithm critically depends on the sampling sequence $(\delta_i)_{i=1}^{k}$. Although the definition of sampling scales does not give an upper bound on the number of sample points, the strategy in Eq. (4) guarantees a constant factor approximation of the minimal number of sample points for a given sampling scale δ, which is $\mathcal{O}(n/(2\delta)^d)$, with d the dimension of the sampling space. For brain core areas, $d \approx 3$, while for cortex areas, $d \approx 2$. Since $2\delta \geq 1$ we use $\mathcal{O}(n/(2\delta)^2)$ as an upper bound in all practical cases.

Theorem 1 (Running time of the iterative refinement algorithm). *Let $(\delta_i)_{i=0}^{k}$ be the sequence of sampling scales. Let n be the number of voxels in the seed and target regions. Let Area(S_t) and Area(T_t) be the surface areas of the well-connected regions of S and T respectively. Assuming that the difference in connectivity between fibers inside and outside of the target bundle is sufficiently large and that the running time is dominated by the cost of computing geodesics, the total running time of the algorithm is*

$$\mathcal{O}\left(\frac{n^2}{(\delta_0)^4} + Area(S_t)Area(T_t)\sum_{i=0}^{k-1}\frac{\delta_i^2}{(\delta_{i+1})^4}\right). \tag{6}$$

---

[2] To regularize the tessellation near the domain boundary, we add a band of 'dummy' sample points around the sampling domains to avoid 'overstretching' tetrahedra. These samples are not used to construct any tracks and are always labeled '−', and are solely used for the construction of the Delaunay tessellation.

Proof. In the zeroth iteration, we have $\mathcal{O}(n/(2\delta_0)^2)$ sample points in S and T. Since all pairs of seed and target sample points are connected, computation of all geodesics between these pairs has a running time given by $\mathcal{O}(n^2/(2\delta_0)^4)$. At the i-th iteration ($i \geq 0$), the set of active tetrahedra in which we sample points is entirely contained in a $4\delta_i$-wide band around the surface of the well-connected regions. To see this, consider a point $p \in S$ further than $4\delta_i$ away from the boundary ∂S_t. If it were in an active tetrahedron, then at least one vertex q of this tetrahedron must be on the opposite side of the boundary of the well-connected region and thus q must be further than $4\delta_i$ away from p. Then the radius of the sphere circumscribing the tetrahedron is strictly larger than $2\delta_i$, implying that this ball of radius $2\delta_i$ contains at least five sampling points, since the sampling density is δ_i, cf. Eq. (3). Since this is in contradiction with the properties of a Delaunay tessellation, the point p must be in an inactive tetrahedron. Note that near the domain boundary the circumsphere may lie partially outside of the domain, but because we have a band of dummy sample points around the domain, the argument still holds, recall footnote 2. Therefore, the total volume in which we sample points is $\mathcal{O}(\text{Area}(S_t)\delta_i)$, so that the number of points sampled in S at iteration i is $\mathcal{O}(\text{Area}(S_t)\delta_i/\delta_{i+1}^2)$, and similarly for region T. Multiplying the number of samples in S by the number of samples in T and the computational cost per geodesic completes the proof upon summing over all iterations i. □

Further Optimization by Local Searching

We may further exploit the assumption of topographic organization—points close together in the seed region connect to points close together in the target region—by restricting the search for connections to connections that are 'close' to the best connections from the previous iteration. This variant of the algorithm will be referred to as *iterative refinement with local searching.*

Consider a sample point $p \in S$ at iteration $i \geq 1$ together with the tetrahedron from which it was sampled. Note that this tetrahedron is constructed at iteration $i-1$, thus at the coarser scale δ_{i-1}. For each of the vertices v of this tetrahedron determine the track $\gamma_v = \arg\max\{C(\gamma) : \gamma(0) = v\}$, i.e. the track with maximal connectivity among all tracks connecting v to the target region. Then $\gamma_v(1) \in T$ is a vertex of a tetrahedron in the target region. We restrict the search by connecting p only to sample points $q \in T$ which are sampled from tetrahedra that include (at least) one of the $\gamma_v(1)$ as their vertices.

Thus in each iteration after the first, we search around at most 4 vertices for each sample point. If we assume that the sample points are approximately uniformly distributed, then the average degree of each vertex in the Delaunay tessellation is $\mathcal{O}(1)$ [8]. Therefore, for each vertex v we have $\mathcal{O}(1)$ expected tetrahedra that have v as a vertex, which means we compute a constant number of connections for each sample in the seed region. This reduces the total running time to

$$\mathcal{O}\left(\frac{n^2}{(\delta_0)^4} + \text{Area}(S_t)\text{Area}(T_t)\sum_{i=0}^{k-1}\frac{\delta_i}{(\delta_{i+1})^2}\right). \quad (7)$$

Clearly, we need to choose the sequence of sampling scales $(\delta_i)_{i=0}^k$ as a function of n if we wish to obtain a theoretical speedup. For example, choosing $\delta_i \propto \sqrt{n}$ reduces the theoretical bound of the iterative refinement algorithm with local searching to at least $\mathcal{O}(1 + 1/\sqrt{n})$

3 Experiments

Experiments were performed to reconstruct the right Optic Radiation (OR) and the right Frontal Aslant Tract (FAT). The OR connects the lateral geniculate nucleus to the primary visual cortex. It is largely concerned with the transmission of visual information from the retina of the eye to the visual cortex. Due to retinotopy we can safely assume topographic organization to hold within this bundle [9,20,22]. The FAT is a lesser-known white matter bundle whose functional role has recently attracted attention. The FAT connects the lateral inferior frontal gyrus and the motor areas of the medial superior frontal gyrus. In the left hemisphere it is thought to support speech initiation and verbal fluency [2], while in the right hemisphere it is believed to be related to executive function, in particular inhibitory control [2]. While topographic organization in the FAT has, to our knowledge, not been investigated, we conduct experiments on this bundle to demonstrate that the proposed algorithm performs well on bundles whose topographic fiber organization is not well-known. Furthermore, the FAT is of clinical interest to a collaborating neurosurgeon.

We use DWI data and segmentation of seed and target regions from a patient with a glioma, kindly provided by a collaborating neurosurgeon. The number of voxels in the start and end regions of the OR are 23 and 466. The number of voxels for the FAT are 1774 and 702. The tumor is located in the left hemisphere, near the boundary of the frontal and occipital lobes. The right OR is located far away from the tumor and should thus be unaffected. One of the endpoint regions of the right FAT, consisting of the motor areas of the medial superior frontal gyrus, is adjacent to the tumor and is slightly distorted due to a mass effect. The DWI scan was acquired using a Philips Achieva 3T MRI scanner ($b = 1500$, 50 diffusion directions, 6 $b = 0$ images, 2 mm isotropic voxel size).

We apply the iterative refinement algorithm using varying sampling sequences starting at increasingly coarser scales, viz. the sequences $(\delta_0, \delta_0 - 1, \ldots, 1, 0.5)$ for $\delta_0 = 1, ..., \delta_{0,\max}$. Since we do not use sampling scales below 0.5, we do not sample at a sub-voxel resolution and can simply use voxel center points as samples. This restriction is not inherent in the algorithm and sampling scales below 0.5 can be used by sampling more finely within voxels. When reconstructing the FAT, we use $\delta_{0,\max} = 6$, the coarsest scale for which seed and target regions allow proper sampling. The seed region corresponding to the lateral geniculate nucleus is considerably smaller than those of the FAT, limiting the coarsest scale for the OR to $\delta_{0,\max} = 3$. The total number of computed tracks over all scales combined serves as a measure for the computational load. We also consider $\delta_0 = 0.5$, which forces the algorithm to connect all n^2 voxels in seed and target regions, and serves as a brute-force baseline comparison for sampling sequences with $\delta_0 \geq 1$. We set $g = \mathrm{Adj}(D)$ in Eq. (1) and the connectivity threshold in Eq. (5) to $\lambda = 0.76$ for the FAT and $\lambda = 0.97$ for the OR. These ad hoc thresholds merely serve to illustrate the feasibility of the proposed algorithm, but may be reconsidered.

Figure 1 illustrates the results of the iterative refinement algorithm on the OR with the aforementioned sequences of sampling scales. In all cases the bundle matches very well with the brute force bundle, differing almost exclusively in the number of tracks found, which decreases with increasing initial sampling scale.

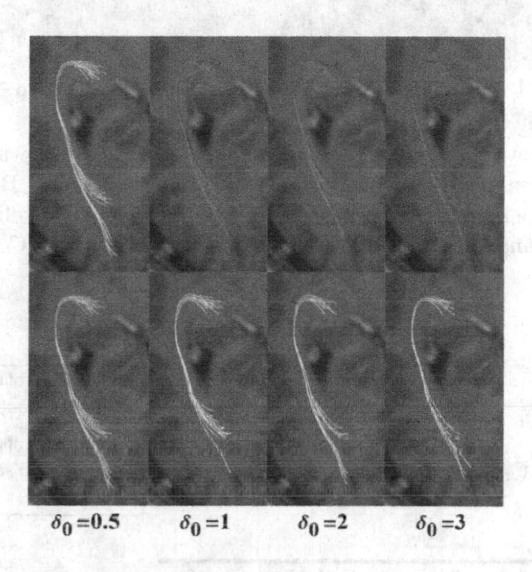

Fig. 1. Axial view of right OR results from the brute force algorithm and the iterative refinement algorithm with various sampling sequences. Brute force results, given by $\delta_0 = 0.5$, are shown in green. Results from the iterative refinement algorithm *without* local searching in red, and *with* local searching in cyan. (Color figure online)

Figure 2 illustrates the results of the iterative refinement algorithm on the FAT with the aforementioned sequences of sampling scales. In all cases a fairly coherent bundle is found, although some seemingly spurious tracks are also present. Such tracks arise as (local) maxima of the connectivity measure (2) and are found by virtue of the criterion (5). These tracks achieve a high connectivity by partially passing through the Corticospinal Tract (CST), essentially taking a high-connectivity detour. The final number of tracks in the last iteration differs for each experiment, showing that the choice of sampling sequence influences the reconstruction of the boundaries ∂S_t and ∂T_t, and thus of the bundle. In particular, we see that for $\delta_{0,\max} = 6$ the resulting bundle is very sparse. We also see that the results from the algorithm with local searching show slightly more 'detour' tracks and are in some cases more sparse than those without local searching. The core bundle is found in all cases.

Figure 3 shows the computational gain when applying the iterative refinement algorithm, both *without and with* local searching, to the right OR and right

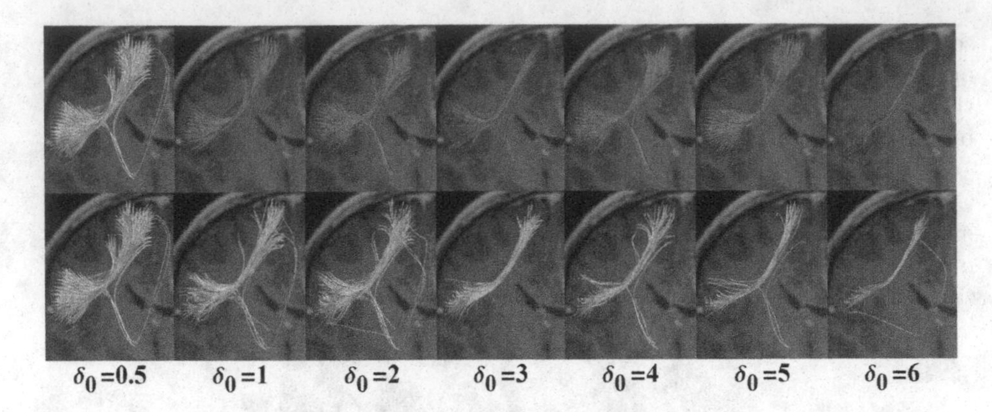

Fig. 2. Coronal view of right FAT results from the brute force algorithm and the iterative refinement algorithm with various sampling sequences. Brute force results, given by $\delta_0 = 0.5$, are shown in green. Results from the iterative refinement algorithm *without* local searching in red, and *with* local searching in cyan. (Color figure online)

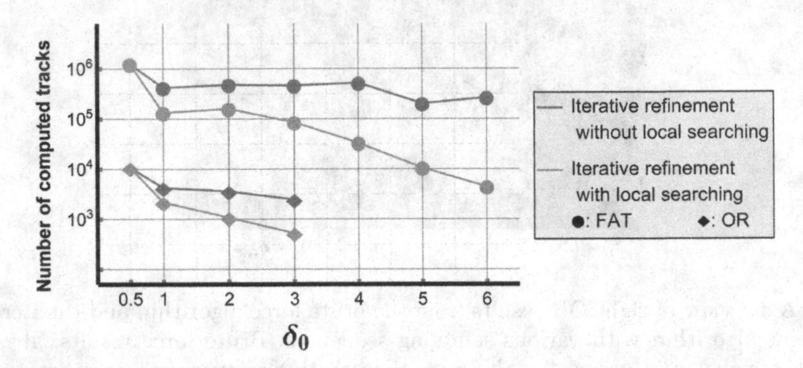

Fig. 3. Total number of geodesic computations in the iterative refinement algorithm without and with local searching applied to the right OR and FAT, as function of the coarsest sampling scale δ_0. The algorithm without local searching achieves a speedup of about one half orders of magnitude, while local searching leads to a speedup of more than one order of magnitude.

FAT for various sampling sequences. The variant with local searching achieves a higher speedup than the variant without local searching, reducing the number of computed tracks by about one order of magnitude on the OR. On the FAT, the variant with local searching attains an even larger speedup of about three orders of magnitude. We stress that the computational effort is measured as the number of computed tracks, not as the physical time it takes for the algorithm to terminate. After all, the iterative refinement algorithm is dependent on the underlying point-to-point tractography algorithm in such a way that improving the latter immediately improves the former proportionally.

In order to illustrate the plausibility of our OR results we compare them in Fig. 4 with the density map of the OR as found by the Juelich probabilistic cytoarchitectonic atlas [1], registered to the structural image of the patient. While clear differences exist, such as the slight 'shortcut' taken by the reconstructed bundle and the lower track density, the agreement is still fairly good. This is particularly true considering that the used techniques cannot be directly compared, since the Juelich atlas was obtained postmortem and we employ in-vivo reconstruction. The presented comparison has therefore a mere illustrative purpose.

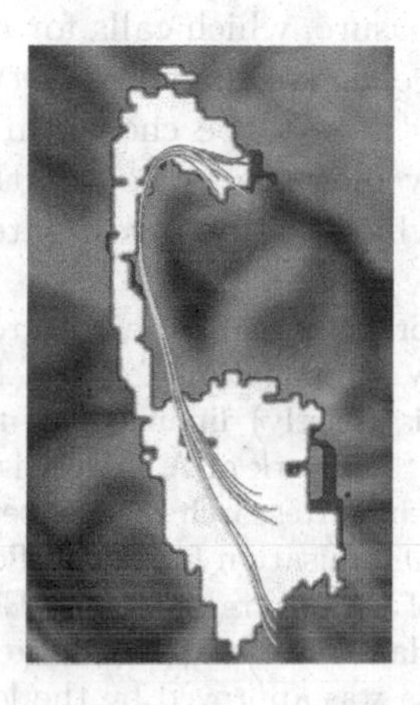

Fig. 4. Axial view of right OR tracks generated by the iterative refinement algorithm on in-vivo DWI without local searching (green) on top of the postmortem OR probabilistic density map as found by the Juelich atlas [1], registered to the structural image of the patient. Yellow white values indicate high probability, red values indicate low probability. (Color figure online)

4 Discussion

We have presented a novel, computationally efficient algorithm for region-to-region tractography. By embedding an existing point-to-point geodesic tractography algorithm into a coarse-to-fine framework, we avoid the computation of all possible n^2 connections between seed and target regions of size n. Our theoretical analysis shows that, under reasonable assumptions, the presented algorithm is significantly more efficient than the quadratic-time solution.

Experiments on a clinical DWI dataset confirm that the proposed iterative refinement algorithm (both with and without local searching) constitutes a step forward towards clinically feasible geodesic tractography, given a proper choice of the sampling scale sequence. The coarsest scale of this sequence should be inferred from the size of the seed and target region. The local search modification provides a significant additional speedup, without seemingly affecting the consistency of the results. The iterative refinement algorithm with local searching is therefore our method of choice for efficient region-to-region geodesic tractography. Its possible use in a clinical setting will be investigated in the context of our collaboration with a neurosurgery department.

The presented algorithm may be adapted in several ways. Firstly, our iterative refinement strategy can be used in combination with any point-to-point tractography algorithm and any connectivity measure. Hence, any improvements in the latter—for example, the development of a connectivity measure that more accurately reflects biological plausibility—would immediately upgrade our results as well. Moreover, we have employed an ad hoc threshold in conjunction with our connectivity measure, which calls for a more rigorously motivated alternative. The difference in effective threshold between the FAT and OR indicates that the threshold may need to be chosen in a bundle-adaptive fashion. Indeed, both the connectivity measure as well as the mentioned threshold are likely to be bundle-specific, which will be investigated in future work.

Acknowledgements. This work is part of the research programme Diffusion MRI Tractography with Uncertainty Propagation for the Neurosurgical Workflow with project number 16338, which is (partly) financed by the Netherlands Organisation for Scientific Research (NWO). The work of A. Fuster is part of the research program of the Foundation for Fundamental Research on Matter (FOM), which is financially supported by the Netherlands Organisation for Scientific Research (NWO). We would like to thank the department of Neurosurgery at the Elisabeth TweeSteden Hospital (ETZ) in Tilburg, The Netherlands, for acquiring the clinical data set used in our experiments. Use of patient data was approved by the local ethics committee (METC Brabant).

References

1. Amunts, K., Mohlberg, H., Bludau, S., Zilles, K.: Julich-brain: a 3D probabilistic atlas of the human brain's cytoarchitecture. Science **369**(6506), 988–992 (2020)
2. Dick A.S., Garic D., Graziano P., Tremblay, P.: The frontal aslant tract (FAT) and its role in speech, language and executive function. Cortex **111**, 148–163 (2019)
3. Astola, L., Florack, L., ter Haar Romeny, B.: Measures for pathway analysis in brain white matter using diffusion tensor images. In: Karssemeijer, N., Lelieveldt, B. (eds.), Proceedings of the Twentieth International Conference on Information Processing in Medical Imaging-IPMI 2007 (Kerkrade, The Netherlands), **4584**, Lecture Notes in Computer Science, 642–649. Springer-Verlag, Berlin (2007). https://doi.org/10.1007/978-3-540-73273-0_53
4. Aydogan, D.B., Shi, Y.: Tracking and validation techniques for topographically organized tractography. NeuroImage **181**, 64–84 (2018)
5. Basser, P.J., Pajevic, S., Pierpaoli, C., Duda, J., Aldroubi, A.: In vivo fiber tractography using DT-MRI data. Magn. Reson. Med. **44**(4), 625–632 (2000)
6. Humphries, C., Liebenthal, E., Binder, J.R.: Tonotopic organization of human auditory cortex. NeuroImage **50**(3), 1202–1211 (2010)
7. Delaunay, B.: Sur la sphère vide. a la mémoire de Georges Voronoï. Bulletin de l'Académie des Sciences de l'URSS. Classe des sciences mathématiques et na. **6**, 793–800 (1934)

8. Dwyer, R.A.: Higher-dimensional Voronoi diagrams in linear expected time. Discrete Comput. Geom. **6**(3), 343–367 (1991)
9. Stephen Engel, G.H.G., Wandell, B.: Retinotopic organization in human visual cortex and the spatial precision of functional MRI. Cerebral Cortex (New York, N.Y. : 1991). **7**, 181–92 (1997)
10. Florack, L., Sengers, R., Meesters, S., Smolders, L., Fuster, A.: Riemann-DTI geodesic tractography revisited. In: Özarslan, E., Schultz, T., Zhang, E., Fuster, A. (eds) Anisotropy Across Fields and Scales. Mathematics and Visualization. Springer, Cham (2021). https://doi.org/10.1007/978-3-030-56215-1_11
11. Fuster, A., Dela Haije, T., Tristán-Vega, A., Plantinga, B., Westin, C.-F., Florack, L.: Adjugate diffusion tensors for geodesic tractography in white matter. J. Math. Imaging Vis. **54**(1), 1–14 (2016)
12. Teofilo, F.G.: Clustering to minimize the maximum intercluster distance. Theoret. Comput. Sci. **38**, 293–306 (1985)
13. Patel, G.H., Kaplan, D.M., Snyder, L.H.: Topographic organization in the brain: searching for general principles. Trends Cogn. Sci. **18**(7), 351–363 (2014)
14. Kuhn, H.W.: The Hungarian method for the assignment problem. Naval Res. Logist. Q. **2**, 83–97 (1955)
15. Lenglet, C., Deriche, R., Faugeras, O.: Inferring white matter geometry from diffusion tensor MRI: application to connectivity mapping. In: Pajdla, T., Matas, J., (eds.), Proceedings of the Eighth European Conference on Computer Vision (Prague, Czech Republic, May 2004), **3021–3024**, Lecture Notes in Computer Science, 127–140. Springer-Verlag, Berlin (2004)
16. Munkres, J.: Algorithms for the assignment and transportation problems. J. Soc. Indus. Appl. Math. **5**(1), 32–38 (1957)
17. O'Donnell, L., Haker, S., Westin, C. F.: New approaches to estimation of white matter connectivity in diffusion tensor MRI: elliptic PDEs and geodesics in a tensor-warped space. In: Dohl, T., Kikinis, R., (eds.), Proceedings of the 5th International Conference on Medical Image Computing and Computer-Assisted Intervention-MICCAI 2002 (Tokyo, Japan, September 25–28 2002), **2488–2489**. Lecture Notes in Computer Science, 459–466. Springer-Verlag, Berlin (2002). https://doi.org/10.1007/3-540-45786-0_57
18. Prados, E., et al.: Control theory and fast marching techniques for brain connectivity mapping. In: Proceedings of the IEEE Computer Society Conference on Computer Vision and Pattern Recognition, New York, USA, June 2006, Vol. 1, pp. 1076–1083. IEEE Computer Society Press (2006)
19. Sengers, R., Fuster, A., Florack, L.: Geodesic tubes for uncertainty quantification in diffusion MRI. In: Sommer, S., Feragen, A., Schnabel, J., Nielsen, M., (eds.), Proceedings of the Twenty-Seventh International Conference on Information Processing in Medical Imaging-IPMI 2021 (Bornholm, Denmark), **12729**, Lecture Notes in Computer Science, 279–290. Springer-Verlag, Berlin (2021). https://doi.org/10.1007/978-3-030-78191-0_22
20. Silver, M., Ress, D., Heeger, D.: Topographic maps of visual spatial attention in human parietal cortex. J. Neurophysiol. **94**, 1358–1371 (2005)
21. Talavage, T.M., Sereno, M.I., Melcher, J.R., Ledden, P.J., Rosen, B.R., Dale, A.M.: Tonotopic organization in human auditory cortex revealed by progressions of frequency sensitivity. J. Neurophysiol. **91**, 1282–1296 (2004)
22. Wandell, B.A., Dumoulin, S.O., Brewer, A.A.: Visual field maps in human cortex. Neuron **56**(2), 366–383 (2007)

Fast Tractography Streamline Search

Etienne St-Onge[1(✉)], Eleftherios Garyfallidis[2], and D. Louis Collins[1]

[1] NeuroImaging and Surgical Technologies Laboratory (NIST), Montreal Neurological Institute (MNI), McGill University, Montreal, QC, Canada
etienne.st-onge@mail.mcgill.ca

[2] Department of Intelligent Systems Engineering, School of Informatics and Computing, Indiana University, Bloomington, USA

Abstract. In this work, a new hierarchical approach is proposed to efficiently search for similar tractography streamlines. The proposed streamline representation enables the use of binary search trees to increase the tractography clustering speed without reducing its accuracy. This hierarchical framework offers an upper bound and a lower bound for the point-wise distance between two streamlines, which guarantees the validity of a proximity search. The resulting approach can be used for fast and accurate clustering of tractography streamlines.

Keywords: Tractography · Streamline · Polyline · White matter bundle

1 Introduction

In classical anatomy, the study of white matter bundles and fascicles connecting different brain regions required dissection. The non-invasive study of these connections has been greatly facilitated by the use of Diffusion Weighted MRI [4,18]. Diffusion Weighted MRI combined with tractography algorithms can be employed investigate the white matter structure [8,17,35].

Streamlines reconstructed from a tractography algorithm are composed of an ordered list of points, depicting white matter pathways. Each streamline is polygonal chain, which is a set of connected line segments (or polyline), with some specific characteristics that depend on the tractography algorithm. For example, most tractography algorithms reconstruct streamlines with a fixed step size (segment length) and a maximum turning angle [2,7,14,30].

Multiple tractography applications require a grouping of similar streamlines for further analysis. Streamlines can be clustered based on similarity and proximity. Numerous algorithms and definitions of distance have been studied to improve accuracy or efficiency of streamlines clustering [11,12,15,24,29,32]. Searching for the nearest streamline in a pre-segmented set of streamlines (called a bundle atlas) can be used to automatically dissect a tractogram [3,13].

In parallel, similar methods have been proposed in the data-mining field to analyze and compare time series data [9,20,22,36]. A multivariate time series

S. Cetin-Karayumak et al. (Eds.): CDMRI 2021, LNCS 13006, pp. 82–95, 2021.
https://doi.org/10.1007/978-3-030-87615-9_8

could also be represented as a polygonal chain. However, current distance measures for tractography streamlines do not directly fit in this framework. Nonetheless, some dimensionality reduction techniques employed for time series can be adapted to an existing streamline distance measure [5,19,36,38].

In this work, we focus on a streamline representation/simplification that conserves important distance properties. The resulting hierarchical representation enables the use of standard binary search trees to increase the clustering speed. In addition, the theoretical upper and lower bounds are used to ensure the accuracy of the proximity search. The resulting formulation can be applied to efficiently compute an exact nearest neighbor (NN) or k-nearest neighbors (KNN) search within a maximum distance.

2 Methods

A tractography streamline $S = [\mathbf{s}_1, ..., \mathbf{s}_m]$ is defined as an ordered series of m points, where each of those points lives in a three-dimensional space $\mathbf{s}_i \in \mathbb{R}^3$, $i \in \{1, ..., m\}$.

2.1 Distance Between Two Points (Vector Norm)

The distance between two points in n-dimensions ($\mathbf{x}, \mathbf{y} \in \mathbb{R}^n$) is generally defined by the Minkowski distance (L^p-norm).

$$\mathrm{dist}_{L^p}(\mathbf{x}, \mathbf{y}) := ||\mathbf{x} - \mathbf{y}||_p = \left(\sum_{j=1}^{n} |x_j - y_j|^p \right)^{\frac{1}{p}} \tag{1}$$

This distance is a generalization of both the Manhattan (L^1) and the Euclidean (L^2) distance. It satisfies the triangle inequality for any $p \geq 1$, resulting in a valid metric. This can be extended to define the maximum norm (L^∞) as $p \to \infty$, dual to the L^1 norm in finite-dimensional spaces.

$$\mathrm{dist}_{L^\infty}(\mathbf{x}, \mathbf{y}) := ||\mathbf{x} - \mathbf{y}||_\infty = \max_{j=1}^{n} (|x_j - y_j|) \tag{2}$$

This research focuses on the L^1 and the L^2 distance. Nonetheless the L^∞ provides some bounding capacity and is sometimes used in binary search trees.

2.2 Distance Between Two Streamlines

The proposed approach will focus on the sum (or average) of L^p-norm to compute the distance between two streamlines ($U = [\mathbf{u}_1, ..., \mathbf{u}_m], W = [\mathbf{w}_1, ..., \mathbf{w}_m]$).

$$\mathrm{dist}_{L^p}(U, W) := \sum_{i=1}^{m} ||\mathbf{u}_i - \mathbf{w}_i||_p = \sum_{i=1}^{m} \left(\sum_{j=1}^{n} |u_{i,j} - w_{i,j}|^p \right)^{\frac{1}{p}} \tag{3}$$

$$\text{mdist}_{L^p}(U, W) := \frac{1}{m}\,\text{dist}_{L^p}(U, W) \tag{4}$$

$$\text{dist}_{\text{MDF}}(U, W) := \min\left(\text{mdist}_{L^2}(U, W),\ \text{mdist}_{L^2}(U, W')\right) \tag{5}$$

The "$\text{mdist}_{L^p}(\cdot,\cdot)$" is employed to compute the average point-wise distance. This is done to normalize the distance by the number of points. When computed for both ascending ($W = [\mathbf{w}_1, ..., \mathbf{w}_m]$) and descending ordered points ($W' = [\mathbf{w}_m, ..., \mathbf{w}_1]$), the average L^2 distance is equivalent to the minimum-average direct flip (MDF) proposed by [11]. The minimum-average direct flip is often used for streamlines clustering, similarity search and registration [13,24]. For tractography, each point is in a tridimensional space, but this measure could be used in higher dimensions. The "$\text{dist}_{L^p}(\cdot,\cdot)$" between two streamlines is depicted in Fig. 1, equivalent to the sum of the norm of directed vectors ($\mathbf{d}_i = \mathbf{u}_i - \mathbf{w}_i$).

2.3 Sum of Norm Properties

In this subsection, a few interesting properties of the sum of L^1 and L^2 are exhibited. These characteristics will be utilized to modify streamline representation while keeping important distance properties. More details are presented in Appendix A.

Remark #1. When using Manhattan (L^1) distance, comparing two lists composed of m n-dimensional points ($\mathbf{u}_i, \mathbf{w}_i \in \mathbb{R}^n$, $i \in \{1, ..., m\}$) is equivalent to computing the L^1 distance between two $m \times n$ dimensional points ($\vec{\mathbf{u}}, \vec{\mathbf{w}} \in \mathbb{R}^{m \times n}$).

$$\text{dist}_{L^1}(U, W) = \sum_{i=1}^{m} \|\mathbf{u}_i - \mathbf{w}_i\|_1 = \|\vec{\mathbf{u}} - \vec{\mathbf{w}}\|_1 \tag{6}$$

Remark #2. The L^1 distance in n-dimensions can be used as an upper and a lower bound obtained from Hölder's inequality.

$$\|\mathbf{x}\|_p \le \|\mathbf{x}\|_q \le n^{(1/q-1/p)}\,\|\mathbf{x}\|_p\,, \text{ for } 0 < q < p \tag{7}$$

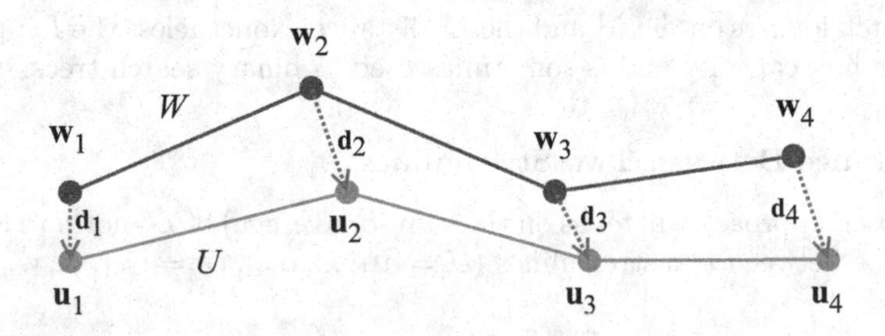

Fig. 1. Pair-wise distance between two tractography streamlines (U, W) from an ordered list of points ($m = 4$), where $\text{dist}_{L^p}(U, W) := \sum_{i=1}^{m} \|\mathbf{u}_i - \mathbf{w}_i\|_p = \sum_{i=1}^{m} \|\mathbf{d}_i\|_p$.

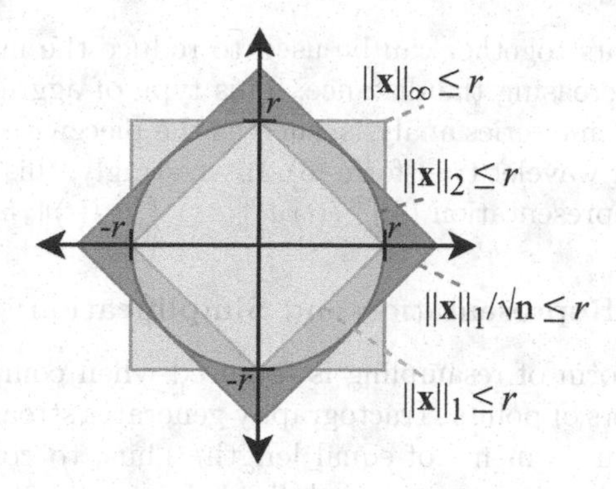

Fig. 2. Illustration of inequality in *Remark #2* (8). In 2D, the Euclidean distance (L^2), displayed in red, is bounded between L^1 and $L^1/\sqrt{2}$.

From this general equation, the euclidean (L^2) distance can be bounded with L^1 with the previous inequality ($q = 1, p = 2$). Figure 2 illustrates this inequality with bounded volume ($\|\mathbf{x}\|_p \leq r$).

$$\frac{1}{\sqrt{n}} \|\mathbf{x}\|_1 \leq \|\mathbf{x}\|_2 \leq \|\mathbf{x}\|_1 \leq \sqrt{n} \|\mathbf{x}\|_2 \tag{8}$$

The sum of distances (p-norm) follow the same rules, since all summed values are positive.

$$\frac{1}{\sqrt{n}} \sum_{i=1}^{m} \|\mathbf{u}_i - \mathbf{w}_i\|_1 \leq \sum_{i=1}^{m} \|\mathbf{u}_i - \mathbf{w}_i\|_2 \leq \sum_{i=1}^{m} \|\mathbf{u}_i - \mathbf{w}_i\|_1 \tag{9}$$

Remark #3. If $\overline{\mathbf{u}} = \frac{1}{m} \sum_{i=1}^{m} \mathbf{u}_i$ is the mean position of a streamline, then the distance between the mean position of two streamlines is always smaller or equal to the average point-wise distance.

$$\text{dist}_{L^p}(\overline{\mathbf{u}}, \overline{\mathbf{w}}) = \left\|\overline{\mathbf{u}} - \overline{\mathbf{w}}\right\|_p \leq \frac{1}{m} \sum_{i=1}^{m} \left\|\mathbf{u}_i - \mathbf{w}_i\right\|_p = \text{mdist}_{L^p}(U, W) \tag{10}$$

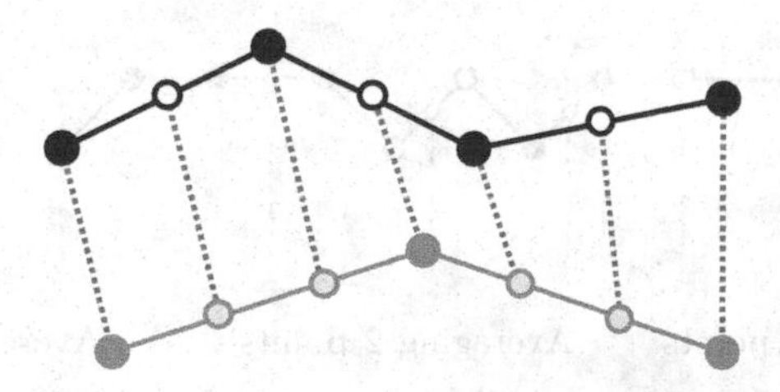

Fig. 3. Uniform resampling of streamlines with an unequal number of points to make both streamlines directly comparable. For equal length segments along each streamline, this can be done by computing the least common multiple of the number of segments.

Thus, averaging points together can be used to reduce the number of points to compare, without increasing the distance. This type of aggregation of points is used extensively for time series analysis, such as the piecewise aggregate approximation [19], or Haar wavelet transform [5]. Interestingly, this remark holds true with a continuous representation ($\| \int_0^1 f(t)\, dt \|_p \leq \int_0^1 \|f(t)\|_p \, dt$, $1 \leq p < \infty$) [6].

2.4 Streamlines Representation and Simplification

Resampling. Some form of resampling is required when comparing streamlines with different numbers of points. Tractography generates streamlines with a fixed step size, resulting in segments of equal length. Thus, to compare streamlines from start to end, each segment can be subdivided according to the least common multiple of the number of segments (see Fig. 3).

Downsampling. Subsampling is often used to reduce streamline complexity, but in the general case, it does not offer any bounding property. Therefore, removing points before the comparison of streamlines can reduce or increase the average distance between them. This is illustrated in Fig. 4-a, where keeping the filled-in points will increase the distance, and keeping the hollow points will decrease it. Thus, some neighbors could be missed when doing a proximity search using subsampled streamlines, resulting in an approximate search.

Averaging. As demonstrated earlier in *Remark #3*, averaging points together never increases the average distance. Therefore, when searching for all similar streamlines in a given radius (r) using averaged points, it is guaranteed that the distance remains inside that radius. Consequently, computing the barycenter, or multiple "sub-barycenters", is an effective way to reduce the number of comparisons (i.e. to reduce dimensionality) for tractography streamline proximity search. This concept of aggregating points will be used to a generate a hierarchical comparison method, similar to a multiscale approach (see Fig. 4-b,c).

2.5 Proposed Hierarchical Streamline Representation

Based on previous remarks, we propose a new hierarchical approach for tractography streamlines proximity search (exact NN or KNN). In this subsection, the

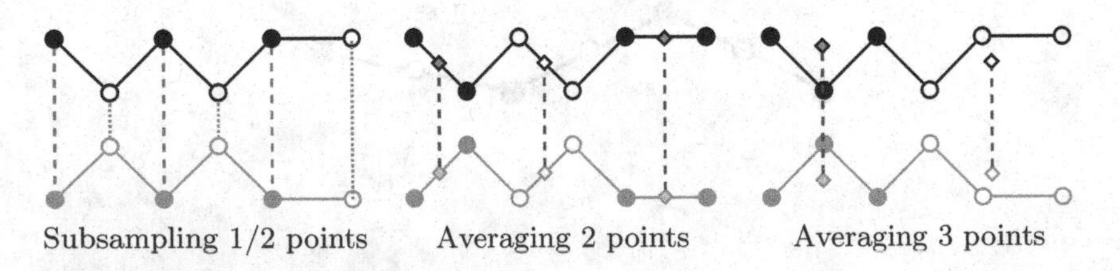

Fig. 4. Subsampling points can reduce (red dotted lines) or increase (gray dashed lines) the overall distance between two streamlines. Averaging points (red dashed lines) does not increase the point-wise distance - it can only reduce it. (Color figure online)

framework is detailed in three procedures: *barycenter binning*, *simplification by averaging*, and *distance refinement*.

Barycenter binning. First, the barycenter of a streamline can be used as an initial proximity search. Because the distance between two barycenters is never greater than the "$\text{mdist}_{L^p}(\cdot,\cdot)$" (see *remark #3*) can be used to limit the proximity search. Coordinates for all barycenters can be grouped together on a regular grid. When searching for all similar streamlines within a specified range (r), only the current grid and its neighbors (within distance r) need to be examined. The binning size can be optimized based on the amount of streamlines and the radius (r) of the proximity search; smaller bins will increase the preprocessing and construction time, but reduce the subsequent search time. Moreover, this barycenter binning provides independent bins, enabling efficient multithreading and the reduction of memory usage. Consequently, each bin can be separated and processed individually, only requiring the current and neighboring bins.

Simplification by averaging. Second, streamline points can be aggregated to create a simplified version with μ mean points. This is done to reduce the number of dimensions when using a binary search tree, thereby "avoiding" the *curse of dimensionality* [23,26,33]. When the number of mean points (μ) is a divisor of the initial number of points (m) the *Remark #3* remains true; otherwise a continuous averaging needs to be done. Afterwards, each streamline's mean points can vectorized in a $\mu \times n$ vector, to employ both *Remarks #1-2.* For the distance between streamlines "$\text{mdist}_{L^p}(\cdot,\cdot)$", with the L^2 norm per points, the searching range need to be increased by the square root of the spatial dimensionality ($\sqrt{n}$). Since the distance between simplified streamlines using μ mean points is always smaller than or equal to the tractography streamline point-wise distance, this aggregation enables the search for all similar streamlines within a certain radius, without missing any streamlines.

Distance refinement. Finally, the resulting similar streamlines with their respective distances, obtained from the proximity search using simplified streamlines, can be refined by computing the full distance (without simplification). When searching for all similar streamlines within a radius (i.e. proximity search), streamlines with a refined distance larger than the desired radius need to be filtered out. For KNN search within a maximum radius, this can be achieved by saving the refined distances to extract the first K streamlines.

3 Experiments

Dataset. For the evaluation, we utilized a single subject (103818) from the Human Connectome Project (HCP) [31]. Tractography streamlines were reconstructed using probabilistic *particle filtering tractography* [14] implemented in *Dipy* [10]. Resulting streamlines were aligned to the MNI space (ICBM 2009a) using ANTs registration [1]. The bundle atlas employed for the experiment is detailed in [13,37]. Streamlines from this atlas were already aligned to the MNI space (ICBM 2009a), which is composed of 30 bundles for all 5 subjects with a total of 255K streamlines. All streamlines were defined with 32 points ($m = 32$), to limits the variability in our testing, and makes the proposed proximity search directly comparable to RecoBundles [13], since RecoBundles downsamples all streamlines to a fixed number of points. For the proposed approach, all streamlines from the bundle atlas were compared with both ascending and descending (flip) order, resulting in 510K streamlines. This makes the "$\text{mdist}_{L^p}(\cdot,\cdot)$" equivalent to RecoBundles' minimum-average direct flip.

Evaluation. The proposed hierarchical streamline search was quantitatively evaluated by measuring the computation time. This computation time did not include any file loading or saving time. The proposed approach was evaluated with and without the *barycenter binning* at various *bin_size* (4 mm, 8 mm, 12 mm). The *simplification by averaging* was compared at different numbers of mean points (2, 4, 8). Without *barycenter binning* and *simplification*, this is equivalent to a brute force search with quadratic time. Multiple sets of streamlines (from the single subject HCP data) were used to vary the total amount of streamlines (500K, 1M, 2M, 4M). The proximity search radius was evaluated from 2 mm to 12 mm, in 2 mm steps. The proximity search was applied to all 30 bundles from the atlas.

The proposed algorithm was also compared to RecoBundles [13] with various number of streamlines. RecoBundles was run with its default parameters from *Scilpy* (v1.1.0): subsampling streamlines to 12 points, pruning distance of 8 mm, and QuickBundles clustering threshold of 12 mm [10,12,28]. It should be noted that the proposed hierarchical proximity search for streamlines is not equivalent to RecoBundles; RecoBundles subsamples streamlines and relies on the QuickBundles clustering algorithm, resulting in an approximate search. However RecoBundles/QuickBundles also prune clusters using an adapted clustering threshold. The goal of this evaluation is to give an idea of the clustering speed of the proposed streamline search method, compared to a state-of-the-art similarity-based clustering method (RecoBundles). Computation times were all measured from a single core on Intel's 2.4 GHz Skylake 6148 processor. The proposed method employs *Scipy* (v1.6.3) cKDTree [34].

4 Results

Figure 5 details the computation time for a tractography streamline proximity search ($\text{mdist}_{L^2}(\cdot,\cdot) \leq 8$ mm) at various numbers of streamlines (500K, 1M, 2M, 4M), searching for similar streamlines in a bundle atlas of 255K streamlines. The resulting performance is comparable to RecoBundles when using both *barycenter binning* and *simplification*. Figure 6 presents the computation time as a function of the search radius (from 2 mm to 12 mm). It can be observed that for smaller radii ($r \leq 4$ mm), *simplification* using only 2 mean points (nb_mpts=2) performs slightly better. For larger radii, using more mean points (nb_mpts=4) results in

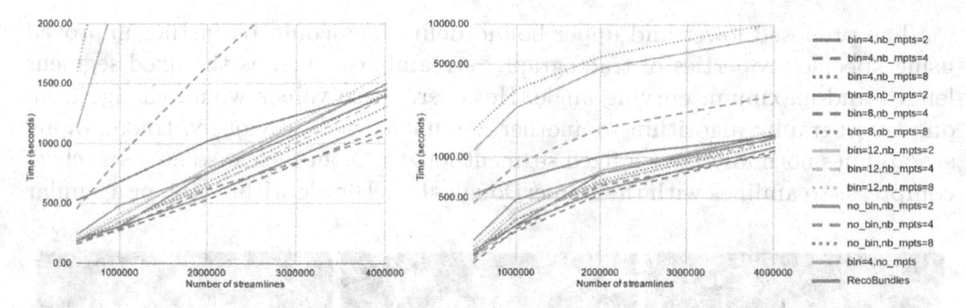

Fig. 5. Streamlines proximity search ($\text{mdist}_{L^2}(\cdot,\cdot) \leq 8\,\text{mm}$) time comparison with multiples parameters, displayed with linear and log scale.

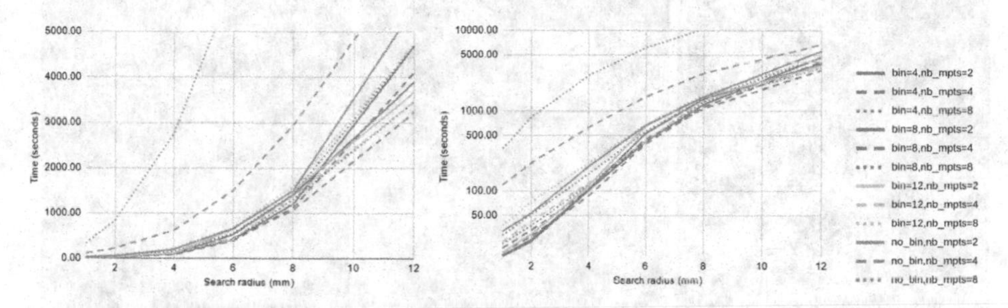

Fig. 6. Time comparison at different search radii ($\text{mdist}_{L^2}(\cdot,\cdot) \leq r$) with 4M streamlines, displayed with linear and log scale.

a smaller computation time. Utilizing more mean points increases the tree search time, but reduces *distance refinement* even more, especially with more streamlines or larger radii. Figure 7 shows streamlines extracted using both the proposed method (radius of 4 mm, 6 mm or 8 mm) and RecoBundles. Both clusters were obtained from the Corticospinal tract (CST) in the bundle atlas. Results for other bundles are displayed in Appendix B.

5 Discussion

Overall, the proposed approach with a *barycenter binning* size between 4 mm and 8 mm and also a *simplification* with 4 mean points results in the fastest computation time. Not using *barycenter binning* results in slower performance (green lines in Figs. 5 and 6). Directly using all streamlines points (32), without *simplification* (orange line in Fig. 5), results in a poor computation time and heavy memory usage for the binary search tree. Bins of 8 mm and 12 mm are not displayed since they were significantly slower than binning at 4 mm.

Depicted in Fig. 7, the proposed framework efficiently and accurately groups similar streamlines and is comparable to existing approaches that use approximate similarity search (RecoBundles). Despite this similarity, this fast streamlines search algorithm could be integrated inside QuickBundles and RecoBundles to further improve their clustering speed when matching bundle centroids.

The proposed lower and upper bound definitions could be further improved using specific properties of tractography streamlines, such as the fixed segment length and maximum curving angle. However, these values would change from one tractography algorithm to another. As mentioned previously, tractography streamlines normally have a fixed segment length. Nonetheless some researchers compress streamlines with the RamerDouglasPeucker algorithm [16], or a similar

Fig. 7. Results of the proximity search for the left Corticospinal tract (CST) of the HCP subject: a) the bundle atlas from [13,37], b) RecoBundles result, c) RecoBundles result (in green) showing in red a few streamlines missing in RecoBundles but present in the proposed technique with a 4 mm search, and in purple, streamlines missing in RecoBundles but present in the proposed technique with a 6 mm search. The proposed proximity search, $\text{mdist}_{L^2}(\cdot,\cdot) \leq r$, using a radius of: d) 4 mm, e) 6 mm, and f) 8 mm. (Color figure online)

variant for tractography [27], to save disk space. Other approaches could be used to reduce the number of points (or dimensions) required when employing a search tree [21,25]. Nevertheless, those dimensionality reduction techniques do not preserve distances nor guarantee any lower/upper limits on distance, resulting in an approximate neighbor search.

6 Conclusion

The proposed framework efficiently and accurately groups similar tractography streamlines without missing any similar streamlines. This method can be used to cluster streamlines into bundles, based on an atlas. The use of *simplification by averaging* (mean points) combined with a binary search tree significantly reduces the query time. Furthermore, *barycenter binning* provides independent bins, enabling efficient multithreading and the reduction of memory usage. Finally, this proposed method guarantees results and is comparable in speed to existing approaches using approximate similarity search.

Acknowledgements. Acknowledgements to Gabrielle Grenier, Maxime Toussaint, Daniel Andrew, Alex Provost for their help and insights. Thanks to the Fonds de recherche du Québec - Nature et technologies (FRQNT), the Canadian Institutes of Health Research (MOP-111169) and the Natural Sciences and Engineering Research Council of Canada (NSERC) for research funding.

Conflict of Interest. We have no conflict of interest to declare.

A Sum of Norm Properties with Detailed Equations

Remark #1. The $\text{dist}_{L^1}(U, W)$ is equivalent to computing the L^1 distance between two $m \times n$ dimensional points ($\vec{\mathbf{u}}, \vec{\mathbf{w}} \in \mathbb{R}^{m \times n}$).

$$\begin{aligned}\text{dist}_{L^1}(U, W) &= \sum_{i=1}^{m} ||\mathbf{u}_i - \mathbf{w}_i||_1 \\ &= \sum_{i=1}^{m}\sum_{j=1}^{n} |u_{i,j} - w_{i,j}| \\ &= ||\vec{\mathbf{u}} - \vec{\mathbf{w}}||_1\end{aligned}$$

Remark #2. The L^1 distance in n-dimensions can be used as an upper and a lower bound the L^2 distance, from Hölder's inequality ($\mathbf{x} \in \mathbb{R}^n$).

$$\begin{aligned}\|\mathbf{x}\|_p \leq \|\mathbf{x}\|_r &\leq n^{(1/r-1/p)} \|\mathbf{x}\|_p && 0 < r < p \\ \|\mathbf{x}\|_2 \leq \|\mathbf{x}\|_1 &\leq \sqrt{n} \|\mathbf{x}\|_2 && r = 1,\ p = 2 \\ \frac{1}{\sqrt{n}} \|\mathbf{x}\|_2 \leq \frac{1}{\sqrt{n}} \|\mathbf{x}\|_1 &\leq \|\mathbf{x}\|_2 && \text{division by } \frac{1}{\sqrt{n}}\end{aligned}$$

Remark #3. The distance between the mean position ($\overline{\mathbf{u}} = \frac{1}{m}\sum_{i=1}^{m} \mathbf{u}_i$) of two streamlines is always smaller or equal to the average point-wise distance. This can be obtained from L^p-norm properties ($1 \leq p < \infty$ $\mathbf{x}, \mathbf{y} \in \mathbb{R}^n$, $\lambda \in \mathbb{R}$).

$$\|\mathbf{x}+\mathbf{y}\|_p \leq \|\mathbf{x}\|_p + \|\mathbf{y}\|_p \qquad \text{triangle inequality} \qquad (11)$$

$$\|\lambda\mathbf{x}\|_p = |\lambda|\,\|\mathbf{y}\|_p \qquad \text{positive homogeneity} \qquad (12)$$

Using $\mathbf{u}_i, \mathbf{w}_i \in \mathbb{R}^n$, $i \in \{1, ..., m\}$, such that $\mathbf{d}_i = \mathbf{u}_i - \mathbf{w}_i$

$$\begin{aligned} \|\overline{\mathbf{u}} - \overline{\mathbf{w}}\|_p &= \left\| \frac{1}{m}\sum_{i=1}^{m} \mathbf{u}_i - \frac{1}{m}\sum_{i=1}^{m} \mathbf{w}_i \right\|_p \\ &= \frac{1}{m}\left\| \sum_{i=1}^{m} (\mathbf{u}_i - \mathbf{w}_i) \right\|_p && \text{from (12)} \\ &\leq \frac{1}{m}\sum_{i=1}^{m} \|\mathbf{u}_i - \mathbf{w}_i\|_p && \text{from (11)} \end{aligned}$$

This can be generalized to curves using the triangle inequality with a Lebesgue integrable function $\| \int_0^1 f(t)\, dt \|_p \leq \int_0^1 \|f(t)\|_p\, dt$, $1 \leq p < \infty$, $f \in \mathcal{L}^1(\mathbb{R})$ [6].

B Streamline Search Comparison

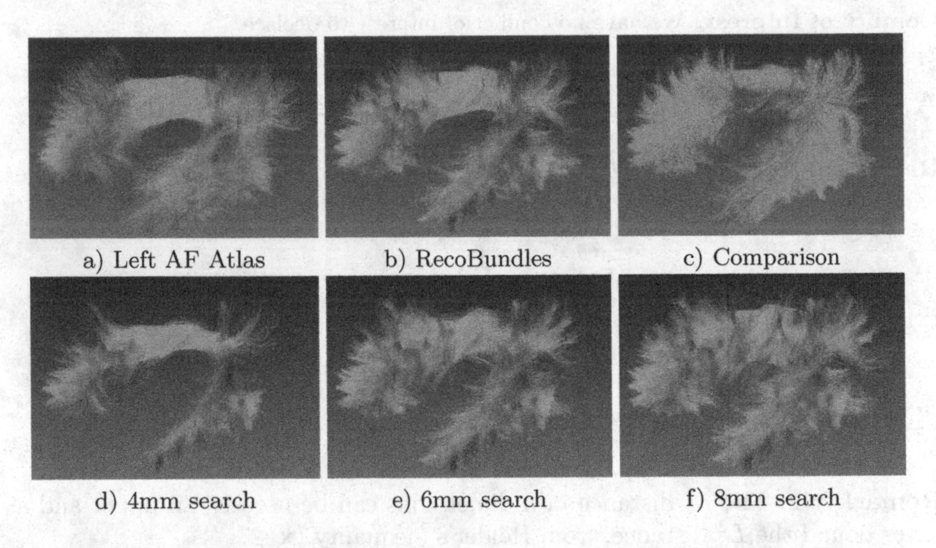

Fig. 8. Results of the proximity search for the left Arcuate Fasciculus (AF): a) the bundle atlas from [13,37], b) RecoBundles result, c) RecoBundles result (in green) showing in red a few streamlines missing in RecoBundles but present in the proposed technique with a 4 mm search, and in purple, streamlines missing in RecoBundles but present in the proposed technique with a 6 mm search. The proposed proximity search, $\text{mdist}_{L^2}(\cdot,\cdot) \leq r$, using a radius of: d) 4 mm, e) 6 mm, and f) 8 mm. (Color figure online)

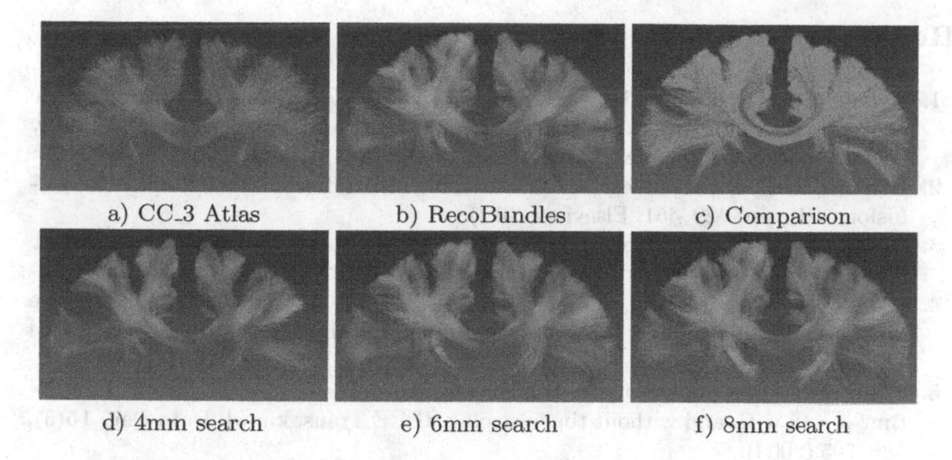

Fig. 9. Results of the proximity search for the central portion of the Corpus Callosum (CC_3): a) the bundle atlas from [13,37], b) RecoBundles result, c) RecoBundles result (in green) showing in red a few streamlines missing in RecoBundles but present in the proposed technique with a 4 mm search, and in purple, streamlines missing in RecoBundles but present in the proposed technique with a 6 mm search. The proposed proximity search, $\text{mdist}_{L^2}(\cdot, \cdot) \leq r$, using a radius of: d) 4 mm, e) 6 mm, and f) 8 mm. (Color figure online)

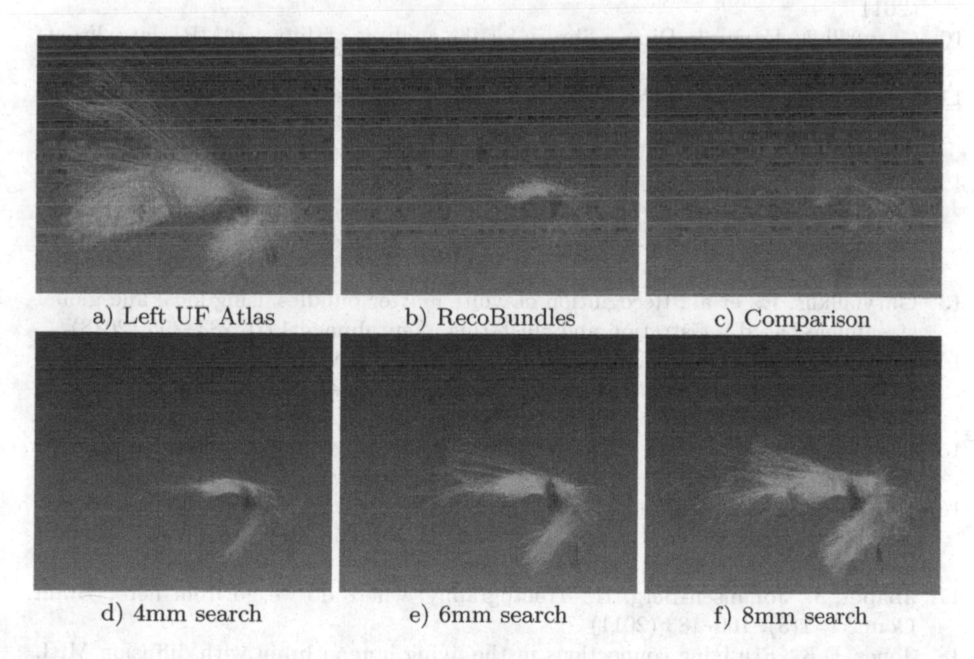

Fig. 10. Results of the proximity search for the left Uncinate Fasciculus (UF): a) the bundle atlas from [13,37], b) RecoBundles result, c) RecoBundles result (in green) showing in red a few streamlines missing in RecoBundles but present in the proposed technique with a 4 mm search, and in purple, streamlines missing in RecoBundles but present in the proposed technique with a 6 mm search. The proposed proximity search, $\text{mdist}_{L^2}(\cdot, \cdot) \leq r$, using a radius of: d) 4 mm, e) 6 mm, and f) 8 mm. (Color figure online)

References

1. Avants, B.B., Epstein, C.L., Grossman, M., Gee, J.C.: Symmetric diffeomorphic image registration with cross-correlation: evaluating automated labeling of elderly and neurodegenerative brain. Med. Image Anal. **12**(1), 26–41 (2008)
2. Behrens, T.E., Sotiropoulos, S.N., Jbabdi, S.: MR diffusion tractography. In: Diffusion MRI, pp. 429–451. Elsevier (2014)
3. Bertò, G., et al.: Classifyber, a robust streamline-based linear classifier for white matter bundle segmentation. NeuroImage **224**, 117402 (2021)
4. Catani, M., Howard, R.J., Pajevic, S., Jones, D.K.: Virtual in vivo interactive dissection of white matter fasciculi in the human brain. Neuroimage **17**(1), 77–94 (2002)
5. Chan, F.P., Fu, A.C., Yu, C.: HAAR wavelets for efficient similarity search of time-series: with and without time warping. IEEE Trans. knowl. Data Eng. **15**(3), 686–705 (2003)
6. Cohn, D.L.: Measure Theory. Springer, Cham (2013)
7. Côté, M.A., Girard, G., Boré, A., Garyfallidis, E., Houde, J.C., Descoteaux, M.: Tractometer: towards validation of tractography pipelines. Med. Image Anal. **17**(7), 844–857 (2013)
8. Descoteaux, M.: High angular resolution diffusion imaging (HARDI). In: Wiley Encyclopedia of Electrical and Electronics Engineering, pp. 1–25 (2015)
9. Fu, T.: A review on time series data mining. Eng. Appl. Artif. Intell. **24**(1), 164–181 (2011)
10. Garyfallidis, E., et al.: Dipy, a library for the analysis of diffusion MRI data. Front. Neuroinf. **21**, 8 (2014)
11. Garyfallidis, E., Brett, M., Correia, M.M., Williams, G.B., Nimmo-Smith, I.: Quickbundles, a method for tractography simplification. Front. Neurosci. **6**, 175 (2012)
12. Garyfallidis, E., Côté, M.A., Rheault, F., Descoteaux, M.: Quickbundlesx: sequential clustering of millions of streamlines in multiple levels of detail at record execution time. In: 24th International Society of Magnetic Resonance in Medicine (ISMRM) (2016)
13. Garyfallidis, E., et al.: Recognition of white matter bundles using local and global streamline-based registration and clustering. NeuroImage **170**, 283–295 (2018)
14. Girard, G., Whittingstall, K., Deriche, R., Descoteaux, M.: Towards quantitative connectivity analysis: reducing tractography biases. Neuroimage **98**, 266–278 (2014)
15. Guevara, P., et al.: Robust clustering of massive tractography datasets. Neuroimage **54**(3), 1975–1993 (2011)
16. Hershberger, J.E., Snoeyink, J.: Speeding up the Douglas-Peucker line-simplification algorithm. University of British Columbia, Department of Computer Science Vancouver, BC (1992)
17. Jbabdi, S., Johansen-Berg, H.: Tractography: where do we go from here? Brain Connect. **1**(3), 169–183 (2011)
18. Jones, D.K.: Studying connections in the living human brain with diffusion MRI. Cortex **44**(8), 936–952 (2008)
19. Keogh, E., Chakrabarti, K., Pazzani, M., Mehrotra, S.: Dimensionality reduction for fast similarity search in large time series databases. Knowl. Inf. Syst. **3**(3), 263–286 (2001)

20. Kotsakos, D., Trajcevski, G., Gunopulos, D., Aggarwal, C.C.: Time-series data clustering. (2013)
21. Legarreta, J.H., et al.: Tractography filtering using autoencoders (2020). arXiv preprint: arXiv:2010.04007
22. Liao, T.W.: Clustering of time series dataa survey. Pattern Recogn. **38**(11), 1857–1874 (2005)
23. Marimont, R., Shapiro, M.: Nearest neighbor searches and the curse of dimensionality. IMA J. Appl. Math. **24**(1), 59–70 (1979)
24. Olivetti, E., Berto, G., Gori, P., Sharmin, N., Avesani, P.: Comparison of distances for supervised segmentation of white matter tractography. In: 2017 International Workshop on Pattern Recognition in Neuroimaging (PRNI), pp. 1–4. IEEE (2017)
25. Olivetti, E., Nguyen, T.B., Garyfallidis, E.: The approximation of the dissimilarity projection. In: 2012 Second International Workshop on Pattern Recognition in NeuroImaging, pp. 85–88. IEEE (2012)
26. Pestov, V.: Is the k-NN classifier in high dimensions affected by the curse of dimensionality? Comput. Math. Appl. **65**(10), 1427–1437 (2013)
27. Presseau, C., Jodoin, P.M., Houde, J.C., Descoteaux, M.: A new compression format for fiber tracking datasets. NeuroImage **109**, 73–83 (2015)
28. Rheault, F.: Analyse et reconstruction de faisceaux de la matière blanche. Computer Science. Université de Sherbrooke (2020)
29. Siless, V., Medina, S., Varoquaux, G., Thirion, B.: A comparison of metrics and algorithms for fiber clustering. In: 2013 International Workshop on Pattern Recognition in Neuroimaging, pp. 190–193. IEEE (2013)
30. Tournier, J.D., Calamante, F., Connelly, A.: Mrtrix: diffusion tractography in crossing fiber regions. Int. J. Imaging Syst. Technol. **22**(1), 53–66 (2012)
31. Van Essen, D.C., et al.: The WU-MINN human connectome project: an overview. Neuroimage **80**, 62–79 (2013)
32. Vázquez, A., et al.: Ffclust: fast fiber clustering for large tractography datasets for a detailed study of brain connectivity. NeuroImage **220**, 117070 (2020)
33. Verleysen, M., François, D.: The curse of dimensionality in data mining and time series prediction. In: Cabestany, J., Prieto, A., Sandoval, F. (eds.) Computational Intelligence and Bioinspired Systems. IWANN 2005. Lecture Notes in Computer Science, **3512**, 758–770. Springer, Berlin, Heidelberg (2005). https://doi.org/10.1007/11494669_93
34. Virtanen, P., et al.: Scipy 1.0: fundamental algorithms for scientific computing in python. Nat. Methods **17**(3), 261–272 (2020)
35. Wakana, S., et al.: Reproducibility of quantitative tractography methods applied to cerebral white matter. Neuroimage **36**(3), 630–644 (2007)
36. Wang, X., Mueen, A., Ding, H., Trajcevski, G., Scheuermann, P., Keogh, E.: Experimental comparison of representation methods and distance measures for time series data. Data Mining Knowl. Disc. **26**(2), 275–309 (2013)
37. Yeh, F.C., et al.: Population-averaged atlas of the macroscale human structural connectome and its network topology. NeuroImage **178**, 57–68 (2018)
38. Yi, B.K., Faloutsos, C.: Fast time sequence indexing for arbitrary LP norms (2000)

Alignment of Tractography Streamlines Using Deformation Transfer via Parallel Transport

Andrew Lizarraga[1(✉)], David Lee[1,3], Antoni Kubicki[1], Ashish Sahib[1], Elvis Nunez[1,2], Katherine Narr[1,4], and Shantanu H. Joshi[1,3]

[1] Ahmanson Lovelace Brain Mapping Center, Department of Neurology, UCLA, Los Angeles, USA
{AndrewLizarraga,dalee,knarr}@mednet.ucla.edu, {akubicki,asahib,elvis.nunez}@ucla.edu, s.joshi@g.ucla.edu

[2] Department of Electrical and Computer Engineering, UCLA, Los Angeles, USA

[3] Department of Bioengineering, UCLA, Los Angeles, USA

[4] Department of Psychiatry and Biobehavioral Sciences, UCLA, Los Angeles, USA

Abstract. We present a geometric framework for aligning white matter fiber tracts. By registering fiber tracts between brains, one expects to see overlap of anatomical structures that often provide meaningful comparisons across subjects. However, the geometry of white matter tracts is highly heterogeneous, and finding direct tract-correspondence across multiple individuals remains a challenging problem. We present a novel deformation metric between tracts that allows one to compare tracts while simultaneously obtaining a registration. To accomplish this, fiber tracts are represented by an intrinsic mean along with the deformation fields represented by tangent vectors from the mean. In this setting, one can determine a parallel transport between tracts and then register corresponding tangent vectors. We present the results of bundle alignment on a population of 43 healthy adult subjects.

1 Introduction

There have been numerous approaches to overcome the difficulty of inter-subject comparison of white matter fiber tracts. Glozman et al. presented a compact geometric model of white matter fibers composed of a center-line and groups of point clouds along the length of the center-line to drive registration and analysis across the subjects [5]. While this could potentially provide an efficient and effective representation and registration, it still may not fully account for the rich geometry within the fiber streamlines. Zhang et al. noted that statistical methods tend to disregard the rich geometry of fiber bundles and proposed to model fibers as a non-parametric Bayesian process which captures the overall geometry and allows for statistical analysis [21]. However, the results in this model are not

S. Cetin-Karayumak et al. (Eds.): CDMRI 2021, LNCS 13006, pp. 96–105, 2021.
https://doi.org/10.1007/978-3-030-87615-9_9

necessarily deterministic. Durrelman et al. proposed an elegant solution by representing the streamlines as currents embedded in the vector fields. This allows estimation of diffeomorphic transformations between two sets of fiber bundles thus capturing the geometric variability across the sets of streamlines [2]. Garyfallidis et al. introduced a framework for streamline-based registration of two fiber tracts without introducing simplified models, and instead solved for a linear transformation that minimizes the cost function of distances between a pair of streamlines [3]. Prasad et al. proposed a maximum density path representation for capturing geometric information of bundles and showed the benefit of automatically segmenting bundles into regions of interest by virtue of the density paths respecting geometrically distinguished portions of the bundle [14]. Another approach by O'Donnell et al. has used distributions of fibers for unbiased group registration of tractography representations [13]. A recent approach by Zhang et al. presented a rich and detailed set of white matter fiber atlas, which allowed direct registration of whole brain white matter fibers, as well as clustering into major fiber bundles [20].

1.1 Contributions of the Paper

In this paper, we rely on parameterized representations of fiber streamlines in an appropriate geometric space with the Fisher-Rao metric [15], along with a low dimensional tangent vector representation of the fibers. This low-dimensional representation takes inspiration from the tangent space principal component analysis developed by Joshi et al. [11]. We propose using this representation, as it reduces the problem of registration to a problem of parallel transport and optimal rotation of the low-dimensional tangent vectors.

In this paper the following contributions are made: i) a new representation that encodes the underlying geometry (mean) of the fiber bundle along with low-dimensional representations of individual fibers about the mean, ii) a new metric in the product space of the mean and the tangent space projections, and finally iii) an alignment method for registering the means and low-dimensional tangent vectors. We show that this process of registration corresponds to tract alignment in the original space. We present the results of bundle alignment on $N = 43$ healthy controls and show significantly lower discrepancy in matching streamlines compared to standard rigid-alignment.

2 Methods

2.1 Preliminary Background

The coordinates of a white matter fiber tract with N fibers can be represented by a set of continuous functions in a Hilbert space, $\mathcal{F} \equiv \{(f_1, f_2, \dots, f_N) \mid f_i \in \mathbb{L}^2([0,1], \mathbb{R}^3)\}$. We sought for a bundle representation that is general and encodes the shape variability of the fiber bundle around the mean. While there are several choices for the shape representations of $\{f_i\}$ including the well-known current-based representations, in this paper we use parameterized curves. Specifically, we use the square-root velocity function (SRVF) to represent the geometry of the fibers [9,10,17]. Each fiber f_i is represented by a SRVF map $f_i \mapsto q_i$, where $q_i = \dot{f}_i/\sqrt{||\dot{f}_i||}$, and imposing the constraint $\int_{[0,1]} \langle q(t), q(t)\rangle \, dt = 1$ ensures that our shape representations are invariant to translation and scaling. We denote the space of such shapes as $\mathcal{S}$. The tangent space of shapes $T_q(\mathcal{S})$ under this representation is endowed by a Fisher-Rao metric which is the standard $\mathbb{L}^2$ inner product invariant under domain reparameterizations of q_i.

2.2 Fiber Tract Representation

We represent the shape of the fiber tract as a pair $\mathcal{B} \equiv (\beta_\mu, \mathcal{A})$, where β_μ is the mean shape of $\{q_i\}$, and $\mathcal{A}$ is a low-dimensional projection of the tangent vectors from the mean shape β_μ in the direction of the shapes q_i. The first component of this representation is computed as the Kärcher mean [12],

$$\beta_\mu = \arg\min_q \frac{1}{N} \sum_{i=1}^{N} \arg\min_{O,\gamma} \left\| q - \sqrt{\dot{\gamma}} \mathcal{O} q_i(\gamma) \right\|^2, \tag{1}$$

where $\mathcal{O} \in SO(3)$, and $\gamma : [0,1] \to [0,1]$ is a reparameterization function.

To derive the second component, we consider a set of tangent vectors $\{v_i \mid v_i \in T_{\beta_\mu}(\mathcal{S}), i = 1, \ldots, N\}$, such that the exponential maps $\exp_{\beta_\mu}(v_i) = q_i$ recover the set of $\{q_i\}$. This map is defined as $\exp_{\beta_\mu}(v_i) = \beta_\mu \cos\alpha + f \sin\alpha$, where $\cos\,\alpha = \langle \beta_\mu, q_i\rangle$ and the initial tangent vector is given by $f = q_i - \langle \beta_\mu, q_2\rangle$. While there are several choices of basis for $T_{\beta_\mu}(\mathcal{S})$, in this work we adopt a Fourier basis. We denote the basis elements as $\{g_k\}, k = 1, \ldots, N$ and project them on $T_{\beta_\mu}(\mathcal{S})$ as $G \equiv \{\tilde{g}_k\}$, where $\tilde{g}_k = g_k - \langle g_k, \beta_\mu\rangle \beta_\mu$, $k = 1, \ldots, N$. Then the second component of the bundle representation $\mathcal{A}$ is given by

$$\mathcal{A} = \langle v_i, g_k\rangle, i = 1, \ldots, N, k = 1, \ldots, N \tag{2}$$

Finally, the fiber tract is represented by the pair of functions,

$$\mathcal{B} \equiv (\beta_\mu, \mathcal{A}),\ \beta_\mu \in \mathcal{S},\ \mathcal{A} \in \mathbb{L}^2([0,1], \mathbb{R}^{N\times N}). \tag{3}$$

Note that the term $\mathcal{A}$ can be interpreted as a coefficient matrix, with the i-th row as the set of Fourier coefficients of the i-th tangent vector, and we treat these coefficients as shape features. Additionally, the rows of $\mathcal{A}$ are in one-to-one correspondence with the tangent vectors. As a consequence, one can change the direction and scale of the tangent vectors by adjusting these coefficients. Since we treat an entire tract as its mean paired with its low-dimensional tangent vectors, the most natural deformation to perform on the $\mathcal{A}$ while respecting the mean are rotational actions in $SO(N)$.

2.3 Approach to the Registration Problem

We provide a rationale to our approach in tract registration: Note that computing the Fisher-Rao metric between Kärcher means, $\inf_\gamma ||\beta_\mu^i - \sqrt{\dot{\gamma}}\beta_\mu^j(\gamma)||$, gives rise to a geodesic between the respective means. This warp is used to register the means between two tracts, by traversing the geodesic from one mean to the other.

However, if one pursues further, we can parallel transport the tangent vectors in $T_{\beta_\mu^i}(\mathcal{S})$ to the tangent space $T_{\beta_\mu^j}(\mathcal{S})$ along the mean geodesic, which is described in detail in Algorithm1. After transport, successful alignment of the two sets of tangent vectors correspond to a registration between the two tractography streamlines. Given the tangent space is high dimensional, we extract the low dimensional representation of the two sets of tangent vectors as described in Eq. 2. An approximate registration then boils down to finding an appropriate rotation between the low-dimensional tangent spaces. The optimal rotation between two low-dimensional features $\mathcal{A}^1$ and $\mathcal{A}^2$, can be found by taking a singular value decomposition of the product $\mathcal{A}^1\mathcal{A}^{2^T} = R\Sigma Q^T$, where $R, Q \in SO(N)$. Then the optimal rotation matrix $\mathcal{O}$ is given by $\mathcal{O} = RQ^T$ [6] (Fig. 1).

2.4 Fiber Bundle Distance

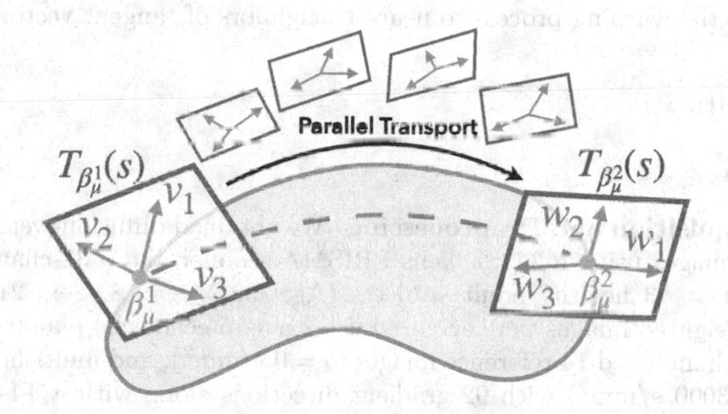

Fig. 1. A visualization of the parallel transport process of one tangent space to another.

We can formalize our approach by stating it as the following metric between two tracts $\mathcal{B}^i = (\beta_\mu^i, \mathcal{A}^i)$, $\mathcal{B}^j = (\beta_\mu^j, \mathcal{A}^j)$ by equipping the induced product-space metric:

$$D(\mathcal{B}^i, \mathcal{B}^j) = \inf_{\gamma, \mathcal{O}} \sqrt{||\beta_\mu^i - \sqrt{\dot{\gamma}}\beta_\mu^j(\gamma)||^2 + ||\mathcal{A}^j - \mathcal{O}(\mathcal{A}^i)||_{fro}{}^2} \quad (4)$$

With some abuse of notation, $\mathcal{A}^i$, is the low dimensional tangent vector representation after parallel transport. The metric on the low-dimensional features is given by the Frobenius metric. By determining the optimal rotation $\mathcal{O} \in SO(N)$ and diffeomorphism $\gamma : [0,1] \to [0,1]$, we get i) a registration between the two tract streamlines and ii) a measure of similarity between tract streamlines. Since

Algorithm 1: Parallel Transport of $\{v_i^2\}$ for the reconstruction of $\mathcal{A}^2$ along a geodesic from β_μ^2 to β_μ^1

Input: $(\beta_\mu^1, \{v_i^1\}, i = 1, \ldots, N)$, $(\beta_\mu^2, \{v_i^1\}, i = 1, \ldots, N)$
Output: Transported tangent vectors $\{\tilde{v}_i^2\}, i = 1, \ldots, N)$
1 Compute a tangent vector w such that $\exp_{\beta_\mu^2}(w) = \beta_\mu^1$
2 Let $l_w = \sqrt{\langle w, w \rangle}$
3 Define a step size k.
4 **for** $\tau \leftarrow 2$ **to** $k - 1$ **do**
5 $\quad q_\tau = \exp_{\beta_\mu}(\frac{w}{k})$
6 $\quad \tilde{v}_i = v_i - \langle v_i, q_\tau \rangle q_\tau$, $i = 1, \ldots, N$
7 $\quad \tilde{v}_i = \tilde{v}_i \frac{l_w}{\|\tilde{v}_i\|}$, $i = 1, \ldots, N$
8 **end**

some shape features may be lost in the low-dimensional setting, we refer to this alignment as a "soft" deformation. Note that a "hard"-alignment can be performed if one then takes the assigned tangent vectors after the soft deformation and applies the warping process to nearest neighbors of tangent vectors.

3 Results

3.1 Data

Image Acquisition and Preprocessing. We obtained diffusion-weighted and structural images using a 3T Siemens PRISMA scanner and a 32-channel head coil from $n = 43$ healthy adult subjects (Age: 32.2 ± 11.8, Sex: 21M/22F). Diffusion-weighted images were acquired using a spin-echo echo planar sequence (EPI), which included 14 reference images ($b = 0$ s/mm^2), and multishell images ($b = 1500, 3000$ s/mm^2) with 92 gradient directions along with a T1-weighted image (voxel size=0.8mm^3). We followed the Human Connectome Project minimal preprocessing pipeline to process the imaging data [4]. T1-weighted structural images were registered to Montreal Neurological Institute (MNI) 152 T1 standard space for anterior commissure - posterior commissure (AC-PC) alignment. 12-degree of freedom (DOF) registration of individual T1 structural image to MNI T1 template was first performed, followed by the 6-DOF registration of individual image to the 12-DOF individual-to-MNI-template registered image. Then, each subject's diffusion weighted image was registered to the structural image using 6-DOF registration [4].

Tractography and Along-Tract Diffusion Measure Extraction. Whole brain tractography was performed with MRtrix3 [18], using multi-shell multi-tissue constrained spherical deconvolution followed by filtering of tractograms [16] to obtain more biologically relevant fibers, ultimately producing 10 million fibers for each subject [8]. Whole brain fiber tracts were segmented into 18 fiber groups; Left (L) and Right (R) Thalamic Radiation (Th Rad), Corticospinal Tract (CST), Cingulate Cingulum (CnCn), Cingulate Hippocampus (CnHp), Superior/Inferior Longitudinal Fasciulus (SLF/ILF), Uncinate (Unc), and Arcuate (Arc), and Corpus Callosum Forceps Major/Minor (CC F Maj/Min) using Automated Fiber Quantification (AFQ) [19]. We selected an arbitrary subject as a representative and performed bundle-wise alignment of all subjects to this template.

Execution Time. Computations of the experiments were performed on a 2.6 GHz 6-Core Intel Core i7 MacBook Pro. The tract-to-tract registration averaged to 33.77 min per registration, with Kärcher mean contributing to approximately 18.48 min of the computational time.

3.2 Within-Tract Fiber Alignment to the Kärcher mean

Prior to performing subject-to-subject tract alignment, we demonstrate the effectiveness of the within-bundle fiber alignment to the mean by observing the profiles of along-tract fractional anisotropy for all fibers in that tract. Figure 2 shows FA profiles after the initial rigid alignment and after undergoing reparameterization resulting from the soft alignment of the L-Arcuate and the L-CST tracts to the template. It is observed that the FA profiles show distinct grouping of features and visually shows reduced variability along the length of the tract, even though the objective function for the soft alignment process did not explicitly incorporate FA as a measure. FA profile alignment is typically used as a qualitative descriptor indicating that a tract is properly aligning to the mean, and we expect to see noisy profiles to form more regular patterns if a registration process is properly aligning the tract. Figure 2 demonstrates how the process of deformation transfer via parallel transport and rotation corresponds to smoother and more regular FA-profiles suggesting that this technique is in fact improving registration over rigid alignment.

3.3 Soft-Alignment of Fiber-Tracts

Figure 3 shows examples of rigid alignment, soft alignment, and hard alignment of L-Cn-Cing, CCF-Major, and R-Uncinate bundles between a subject and the template. The last two columns show the deformation represented by the warping functions γ between the individual fibers of the subject, and the target fibers of the template. Note that the closer these γ functions are to identity function, corresponds to individual fibers in the subject tract being more aligned to the template.

Although the initial rigid alignment brings the subject closer to the template, one can observe discrepancies in the overall geometry of the fibers. Additionally, the warping functions deviate further from the identity indicating that a majority of the fibers are not aligned. The soft alignment improves upon the overlap by achieving an overall agreement of the shapes. We visually see that fibers from

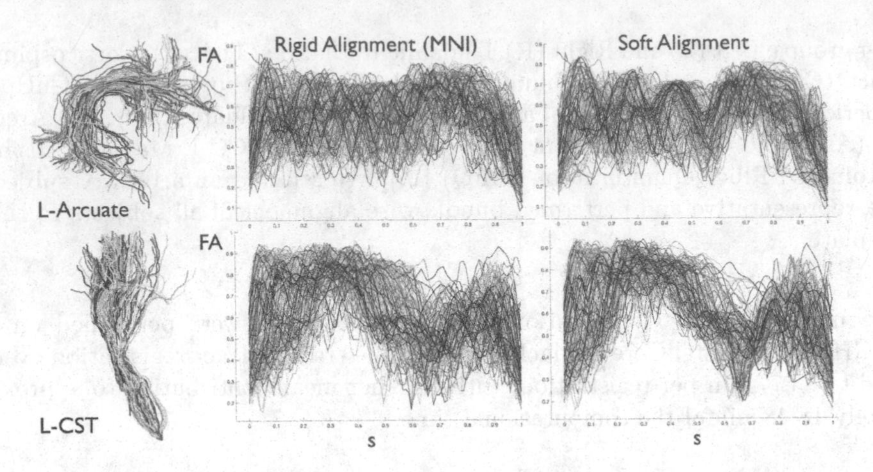

Fig. 2. Left: Fiber tracts of the L-Arcuate and the L-CST. Center: Along-tract FA profiles for all fibers in the respective bundles obtained from the initial (rigid) alignment. Right: Re-parameterized FA profiles after aligning each fiber to the within-tract Kärcher mean.

the subject tract conform to the overall template shape, and that the warping functions have aligned closer to the identity.

The hard alignment improves upon the soft-alignment by further refining over the local details and especially respecting the ends of the fiber tracts. In the case of CCF-Major (middle-row), hard-alignment succeeds in warping the subject fibers to conform to the overall shape of the template. Note, while one can directly pairwise register the fibers between subject and template, the initial pairing of the subject fibers and the template fibers is arbitrary. Thus the warping between such fibers incurs a large cost of reparameterization.

3.4 Tract Point-Set Similarity

There is no direct method for comparing the closeness of tract shapes between the subject and the target, as the rigid-alignment does not provide comparable fiber correspondences between tracts unlike the soft alignment method. To overcome this challenge, we compute the bidirectional Hausdorff distance that can be applied to arbitrary point-sets [7]. This is given by $D_H(\mathcal{B}_1, \mathcal{B}_2) = \max(d_H(\mathcal{B}_1, \mathcal{B}_2), d_H(\mathcal{B}_2, \mathcal{B}_1))$, where $d_H(\mathcal{B}_1, \mathcal{B}_2) = \max_{a \in \mathcal{B}_1} \min_{b \in \mathcal{B}_2} ||a - b||$. The bidirectional Hausdorff distance D_H was computed for each tract for $N = 43$ subjects and was shown to be significantly less after soft alignment for all tracts ($p < 1e-5$) except for the CC F Min ($p = 0.0748$) after correcting for multiple comparisons using FDR [1]. The average Hausdorff distance for CC F Min was still lower for the soft alignment method but did not survive FDR (Fig. 4).

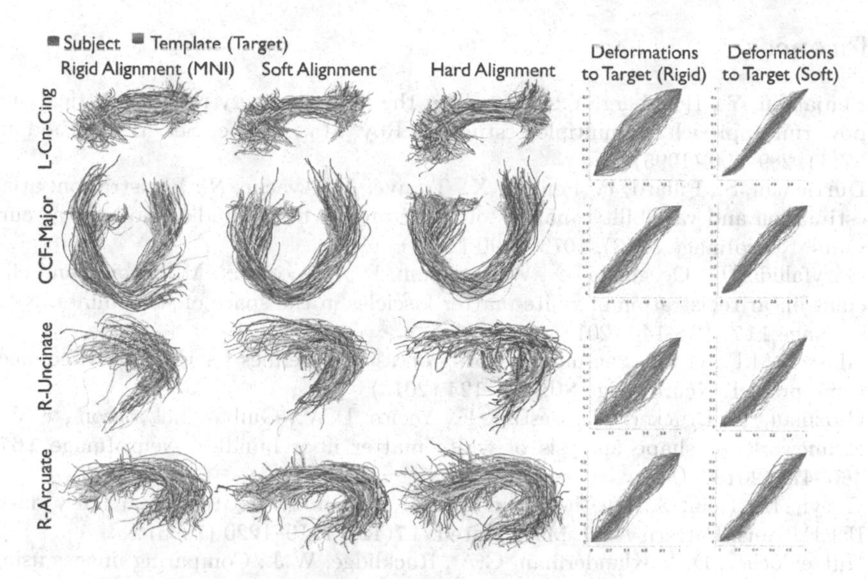

Fig. 3. Comparison of soft and hard alignment between a subject and a template along with warping functions (γ) against rigid alignment in the MNI space.

Bundle	L Th Rad	R Th Rad	L CST	R CST	L CnCn	R CnCn	L CnHp	R CnHp	CC F Maj	CC F Min	L ILF	R ILF	L SLF	R SLF	L Unc	R Unc	L Arc	R Arc
T-stat	43.27	33.75	9.88	22.6	18.5	19.8	3.72	11.0	1.07	1.70	0.72	15.00	10.70	11.52	9.14	8.49	33.23	33.75
P-value	p < 1e-5	p < 1e-5	p < 1e-5	p < 1e-5	p < 1e-5	p < 1e-5	p < 1e-5	p < 1e-5	p < 1e-5	0.0748	p < 1e-5	p < 1e-5	p < 1e-5	p < 1e-5	p < 1e-5	p < 1e-5	p < 1e-5	p < 1e-5

Fig. 4. Comparisons of tract point-set closeness between rigid and soft alignment (significantly smaller) using Hausdorff distance after FDR correction for $N = 43$ subjects. The CC F Min tract (shaded) did not achieve significance.

4 Discussion

We presented a novel framework for soft registration of white matter fiber tracts using a low-dimensional representation that encodes shape deformations. The mechanism of parallel transport and product metric enables an effective computation of tract differences while simultaneously allowing the alignment of tracts. From within-tract fiber-to-mean registration results, we see that the shape alignment of geometrically similar fibers may enhance the features of diffusion measures sampled along their lengths even though the measure (FA) was not explicitly accounted for in the deformation process. This framework is general and will potentially allow statistical shape analysis of general collections of streamlines.

Acknowledgements. This research was partially supported by a fellowship from the NSF NRT Award #1829071 (EN) and the NIH NIAAA (National Institute on Alcohol Abuse and Alcoholism) awards R01-*AA*025653 and R01-*AA*026834 (SHJ). Data acquisition and processing was also supported by NIH/NIMH award U01MH110008.

References

1. Benjamini, Y., Hochberg, Y.: Controlling the false discovery rate: a practical and powerful approach to multiple testing. J. Roy. Statist. Soc. Ser. B (Methodol.) **57**(1), 289–300 (1995)
2. Durrleman, S., Fillard, P., Pennec, X., Trouvé, A., Ayache, N.: Registration, atlas estimation and variability analysis of white matter fiber bundles modeled as currents. NeuroImage **55**(3), 1073–1090 (2011)
3. Garyfallidis, E., Ocegueda, O., Wassermann, D., Descoteaux, M.: Robust and efficient linear registration of white-matter fascicles in the space of streamlines. NeuroImage **117**, 124–140 (2015)
4. Glasser, M.F., et al.: The minimal preprocessing pipelines for the human connectome project. NeuroImage **80**, 105–124 (2013)
5. Glozman, T., Bruckert, L., Pestilli, F., Yecies, D.W., Guibas, L.J., Yeom, K.W.: Framework for shape analysis of white matter fiber bundles. NeuroImage **167**, 466–477 (2018)
6. Goryn, D., Hein, S.: On the estimation of rigid body rotation from noisy data. IEEE Trans. Pattern Anal. Mach. Intell. **17**(12), 1219–1220 (1995)
7. Huttenlocher, D.P., Klanderman, G.A., Rucklidge, W.J.: Comparing images using the hausdorff distance. IEEE Trans. Pattern Anal. Mach. Intell. **15**(9), 850–863 (1993)
8. Jeurissen, B., Tournier, J.D., Dhollander, T., Connelly, A., Sijbers, J.: Multi-tissue constrained spherical deconvolution for improved analysis of multi-shell diffusion MRI data. NeuroImage **103**, 411–426 (2014)
9. Joshi, S.H., Klassen, E., Srivastava, A., Jermyn, I.: A novel representation for Riemannian analysis of elastic curves in R^n. In: IEEE Conference on Computer Vision and Pattern Recognition (CVPR), pp. 1–7. IEEE (2007)
10. Joshi, S.H., Klassen, E., Srivastava, A., Jermyn, I.: Removing shape-preserving transformations in square-root elastic (SRE) framework for shape analysis of curves. In: Energy Minimization Methods in Computer Vision and Pattern Recognition (EMMCVPR), pp. 387–398 (2007)
11. Joshi, S.H., et al.: Statistical shape analysis of the corpus callosum in schizophrenia. NeuroImage **64**, 547–559 (2013)
12. Karcher, H.: Riemannian center of mass and mollifier smoothing. Commun. Pure Appl. Math. **30**, 509–541 (1977)
13. ODonnell, L.J., Wells, W.M., Golby, A.J., Westin, C.F.: Unbiased groupwise registration of white matter tractography. In: Ayache, N., Delingette, H., Golland, P., Mori, K. (eds.) Medical Image Computing and Computer-Assisted Intervention – MICCAI 2012. MICCAI 2012. Lecture Notes in Computer Science, **7512**, 123–130. Springer, Berlin, Heidelberg (2012). https://doi.org/10.1007/978-3-642-33454-2_16
14. Prasad, G., et al.: Automatic clustering and population analysis of white matter tracts using maximum density paths. Neuroimage **97**, 284–295 (2014)
15. Rao, C.: Information and accuracy attainable in estimation of statistical parameters. Bull. Calcutta Math. Soc. **37**, 81–91 (1945)
16. Smith, R.E., Tournier, J.D., Calamante, F., Connelly, A.: Sift2: enabling dense quantitative assessment of brain white matter connectivity using streamlines tractography. Neuroimage **119**, 338–351 (2015)
17. Srivastava, A., Klassen, E., Joshi, S.H., Jermyn, I.H.: Shape analysis of elastic curves in Euclidean spaces. IEEE Trans. Pattern Anal. Mach. Intell. **33**, 1415–1428 (2011)

18. Tournier, J.D., et al.: Mrtrix3: a fast, flexible and open software framework for medical image processing and visualisation. NeuroImage **202**, 116137 (2019)
19. Yeatman, J.D., Dougherty, R.F., Myall, N.J., Wandell, B.A., Feldman, H.M.: Tract profiles of white matter properties: automating fiber-tract quantification. PLOS ONE **7**(11), 1–15 (2012)
20. Zhang, F., Wu, Y., Norton, I., Rigolo, L., Rathi, Y., Makris, N., O'Donnell, L.J.: An anatomically curated fiber clustering white matter atlas for consistent white matter tract parcellation across the lifespan. NeuroImage **179**, 429–447 (2018)
21. Zhang, Z., Descoteaux, M., Dunson, D.B.: Nonparametric bayes models of fiber curves connecting brain regions. J. Am. Statist. Assoc. **114**(528), 1505–1517 (2019)

[illegible]
[illegible]
[illegible]
[illegible]

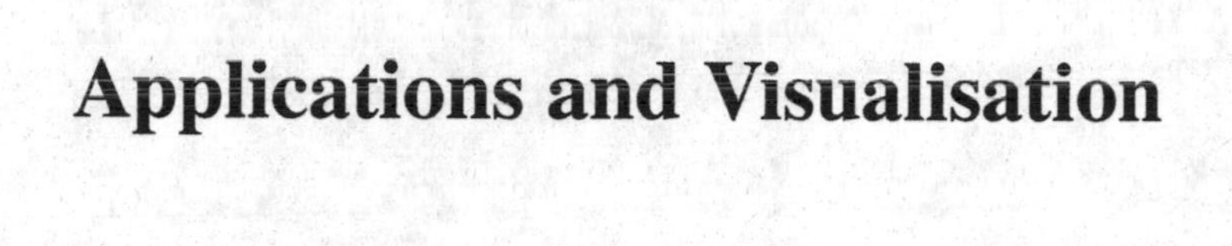

Applications and Visualisation

Diffusion MRI Automated Region of Interest Analysis in Standard Atlas Space versus the Individual's Native Space

Lanya T. Cai(✉), Maria Baida, Jamie Wren-Jarvis, Ioanna Bourla, and Pratik Mukherjee

Department of Radiology & Biomedical Imaging, UCSF, San Francisco, CA, USA
tianhao.cai@ucsf.edu

Abstract. White matter microstructures have been studied most commonly using diffusion tensor imaging (DTI) that models diffusivity in each voxel of diffusion MRI images as a tensor. Classic DTI parameters (e.g., mean diffusivity or MD, fractional anisotropy or FA) derived from the eigenvalues of tensors have been widely used to describe white matter properties. More recently, novel metrics like neurite orientation dispersion and density imaging (NODDI) have broadened the spectrum over which we can both characterize healthy connectivity and investigate pathology. When looking at specific brain regions, previous works combining DTI and NODDI have focused on regions of interest (ROI) analysis where regional masks were generated by mapping known atlas to standard spaces and applied to skeletonized FA maps from tract-based spatial statistics (TBSS). Recent advancement in probabilistic tractography, e.g., the FSL XTRACT toolbox, provides an alternative method of ROI analysis by estimating tract regions in an individual native diffusion space, but the exact advantages and disadvantages compared to using a standard space have not been well documented. In the present study, we perform ROI analysis on DTI and NODDI parameters from diffusion MRI (dMRI) of 39 healthy adults collected from two time points, using both standard-space method ("TBSS ROI analysis") and native-space method ("XTRACT ROI analysis"). We compare the test-retest reliability of these two methods by evaluating the coefficient of variation (C_V) at each time point, the Pearson's correlation (R) between the two time points, and the intra-class correlation coefficient (ICC) between the two time points. With these statistics, we aim to determine the precision of the TBSS ROI analysis and the XTRACT ROI analysis quantitatively in the practice of analyzing a particular dataset. The prospective results will provide a new and general reference for choosing analysis methods in future dMRI studies.

Keywords: DTI · Probabilistic tractography · NODDI · XTRACT

S. Cetin-Karayumak et al. (Eds.): CDMRI 2021, LNCS 13006, pp. 109–120, 2021.
https://doi.org/10.1007/978-3-030-87615-9_10

1 Introduction

Diffusion MRI (dMRI) is a powerful method to study white matter (WM) connectivity non-invasively and in vivo [1,12,13]. Among all parametric models of dMRI data, diffusion tensor imaging (DTI) has been most commonly exploited in characterizing WM microstructural organization. Despite the wide application, DTI alone has inherent limitations as it does not represent the complex WM crossing-fibre configurations [10]. The emerging multi-shell parametric models, e.g., neurite orientation dispersion and density imaging (NODDI), has advanced dMRI research by providing additional aspects of WM properties. Combining DTI and NODDI has shown promise in clinical applications, for example, to identify pathology after mild traumatic brain injury (mTBI) [19,24]. Palacios et al. (2020) used DTI and NODDI to predict both short- and long-term effects of mTBI on WM microstructure, helping to advance mTBI management and treatment. Continuing this line of work is meaningful considering the growing impacts of TBI on public health worldwide [11,14].

In this project, we analyze 2-shell dMRI data of 39 healthy adults from the TRACK-TBI study (Transforming Research and Clinical Knowledge in Traumatic Brain Injury) [23] using DTI and NODDI models, with a specific focus on comparing two alternative methods for Region of Interest (ROI) analysis. Previous works combining DTI and NODDI have used the JHU atlas and tract-based spatial statistics (TBSS) [20] WM ROI analysis [19]. One major limitation is that the standard space does not always capture individual variability in microstructure. For example, characterizing pathological changes in moderate to severe TBI with a standard space is not feasible due to registration challenges. Recent advancement in diffusion tractography provides an alternative method for individualized measures of WM tracts. In particular, the XTRACT package developed by the FSL team is one of the latest pipelines for probabilistic tractography [22]. It enables efficient estimation of an individual's tractography in the native diffusion space. This is critical in studying a broader spectrum of TBI pathology, because patients exhibit a wide range of structural changes that has been challenging to register to and measure in a standard space template.

We aim to compare these two methods by: (1) measuring the variability of ROI analysis results to infer the precision of each method; (2) assessing the test-retest reproducibility to estimate the reliability of each method [15,16]. We expect to see greater precision and reliability in TBSS ROI analysis results, but expect to see greater inter-subject variability in the XTRACT ROI analysis pipeline. The details of the two pipelines as well as the methods for the comparative analysis are described below.

2 Methods

2.1 Participants and Image Acquisition

We analyzed a subset of the multi-site collaboration database of the TRACK-TBI project. This subset contains 39 healthy adults from the control groups from 5 different sites, detailed in Table 1. Using healthy subjects is necessary for reliability studies, because patients may present pathological changes over time. Whole-brain diffusion MRI and T1 were collected from each subject at two time points approximately 6 months apart. Diffusion MRI was performed with multi-slice single-shot spin-echo echo-planar pulse sequence, acquired at b = 1000 (echo time = 94 ms; repetition time = 2900 ms) and 3000 s·mm^{-2} (echo time = 122 ms; repetition time = 3500 ms), both for 64 diffusion-encoding directions (slice thickness = 2.4 mm; matrix= 96 × 96; FOV = 230 mm). Separate b = 0 s·mm^{-2} volumes were acquired in the reversed phase-encoding direction for each diffusion shell with corresponding acquisition parameters to be used for susceptibility distortion correction. Sagittal three-dimensional (3D) inversion recovery fast spoiled gradient-recalled echo T1-weighted images were acquired with 256-mm FOV and 200 contiguous partitions (1.2 mm) at 256 × 256 matrix.

Table 1. Basic information of database in the current study by site.

Site	# of subjects	Age	Scanner model
Baylor College of Medicine	15		Siemens Trio
Harvard Medical School/Massachusetts General Hospital	3		Siemens Skyra
University of Pittsburgh Medical Center	1	38±14	Siemens Prisma
University of Texas at Austin	19		Siemens Skyra
Froedtert and Medical College of Wisconsin	1		GE MR750

2.2 DTI Processing

The FMRIB Software Library (FSL) version 6.0.2 (Oxford, UK) was used for image processing and DTI parameter computation. Susceptibility induced distortions were corrected using FSL's topup [5] on each diffusion shell. FSL's Eddy [6] command was used on the diffusion data to correct for motion and eddy distortions, skull stripping, outlier replacement [4], susceptibility-by-movement [2] and slice-to-volume [3] correction. For each subject, eddy was run once on the b1000 shell and once with concatenated b1000 and b3000 multi-shell data. The Diffusion Toolbox (dtifit) in FSL was used on the individually processed b1000 data to calculate fractional anisotropy (FA) and mean diffusivity (MD). Multi-shell processed data were normalized to their corresponding b0 image to account for differences in echo time. This normalized multi-shell data were used to quantify NODDI parameters including the neurite density index (NDI), orientation dispersion index (ODI) and free water fraction (FISO) with the Accelerated Microstructure Imaging via Convex Optimization (AMICO) Toolbox [9].

2.3 TBSS Analysis

We performed tract-based spatial statistics (TBSS) on the FA map using FSL's TBSS package for each site. The FA maps at both time points for each subject in the site were registered to the FMRIB58 FA template in MNI152 standard space. Each FA map was transformed by combining the non-linear transform to the target FA image and the affine transform from the determined target image to MNI152 space. The registered FA maps were then averaged and thinned to generate a mean FA skeleton to represent the center of all WM tracts. The FA WM skeleton was thresholded to FA > 0.2 to exclude voxels containing gray matter and partial volume effects. Each subject's FA data at both time points were projected onto this mean skeleton to get individual skeletonized FA maps. The voxels within the skeleton were values from the nearest relevant tract center by searching perpendicular to the local skeleton structure for the maximum value in the FA image of the subject. Each subject's FA, MD, NDI, ODI, and FISO maps were then registered and projected onto the WM skeleton. WM tract masks were obtained masking the Johns Hopkins University (JHU) ICBM-DTI-81 White-Matter Labeled Atlas [18] regions to the WM skeleton in MNI152 space. Regional values represented by the average voxel value within the selected JHU WM tract masks are computed for each subject at both time points across all generated DTI and NODDI parameter maps.

2.4 XTRACT Processing and Analysis

XTRACT is a newly developed FSL software package with a library of standardized tractography protocols to derive and extract individualized WM tracts in a subject's native space. Prior to running XTRACT, the data for each time point per subject were prepared following FSL's FMRIB's Diffusion Toolbox (FDT) pipeline which included: brain extraction using the Brain Extraction Tool (BET), fitting the probabilistic crossing fiber model using Bayesian Estimation of Diffusion Parameters Obtained using Sampling Techniques (BEDPOSTX) on eddy corrected multi-shell diffusion data, and linear (FLIRT) and non-linear (FNIRT) registration of diffusion space (first b0 volume of multi-shell data after BET) and structural space (T1 after BET) to standard MNI152 space at 1mm resolution. XTRACT loops through a list of predetermined seeds and corresponding termination criteria warped from standard space to the subject's native space and uses probabilistic tractography (PROBTRACKX2) to define subject specific WM tracts. Tract masks were generated with XTRACT and XTRACT-STATS and used to estimate DTI and NODDI parameter statistics including voxel average from within each masked region, thresholded at 0.1 percent.

2.5 Analysis of Regional Values

From the above pipelines, for each subject we computed regional values for 5 parameters (FA and MD from DTI, NDI, ODI, and FISO from NODDI) from both time points. Due to the computational challenges in probabilistic tractography and the available source image quality, XTRACT occasionally produced empty masks for tracts that were difficult to estimate, resulting in zeros in the estimated regional values. These missing estimations are excluded from later statistical analysis. Among default TBSS and XTRACT regions, we selected 11 out of 48 TBSS regions and 15 out of 42 XTRACT regions for the current study. Our selection was based on availability of equivalent regions in both ROI analysis pipelines (e.g., uncinate fasciculus is available as UNC-L and UNC-R from TBSS regions, and uf-l and uf-r from XTRACT regions). We also excluded regions known to be easily confounded (e.g., fornix regions are easily partial volume averaged with cerebrospinal fluid).

We focused the comparative analysis on test-retest reliability by computing the coefficient of variation (C_V) of each region at each time point across all subjects, the Pearson's correlation (R) between the two time points, and the intra-class correlation coefficient (ICC) between the two time points. The C_V is computed as the ratio of standard deviation and mean, which is the reciprocal of signal-to-noise ratio. In a given comparison, a distribution of regional values with lower C_V has lower variability, therefore indicating a higher precision in the estimation method. Pearson's correlation and intra-class correlation indicate to what extent regional values from the first test were replicated by its counterpart from the second test, with slightly different assumptions on pooling the total variability. The criterion for Pearson's correlation is that the probability of null hypothesis ($R = 0$) is lower than 0.05. The criterion for intra-class correlation is that ICC > 0.6 is considered a good test-retest reliability.

3 Results

First, we will show a direct visualization of regional values by scatter plots. Figure 1 shows regional values of DTI and NODDI parameters in the selected 26 regions that are comparable across both JHU/TBSS and XTRACT methods, labeled at the bottom of the figure.

We have 11 regions from TBSS ROI analysis (with uppercase names) and 15 regions from the XTRACT ROI analysis (with lowercase names). Some of these regions form pairs of bilateral measures. Anatomically comparable regions from the two pipelines are shown in adjacency, although the comparability might not always be obvious from labeled names. Specifically, superior longitudinal fasciculus (SLF) was measured bilaterally as one region in the TBSS pipeline, but as 3 separate components in the XTRACT pipeline. The genu of corpus callosum (GCC) is comparable to the forceps minor ("fmi"), while the splenium of corpus callosum (SCC) is comparable to the forceps major ("fma"). The cingulum - cingulate gyrus (CGC) is comparable to the dorsal cingulum ("cbd").

Importantly, even though these regions are considered closely relevant, we should keep in mind that regional values from the TBSS pipeline are based on skeletonized 2D segments of tracts, whereas the counterparts from XTRACT are based on estimated 3D whole-tracts. This fundamental difference between the

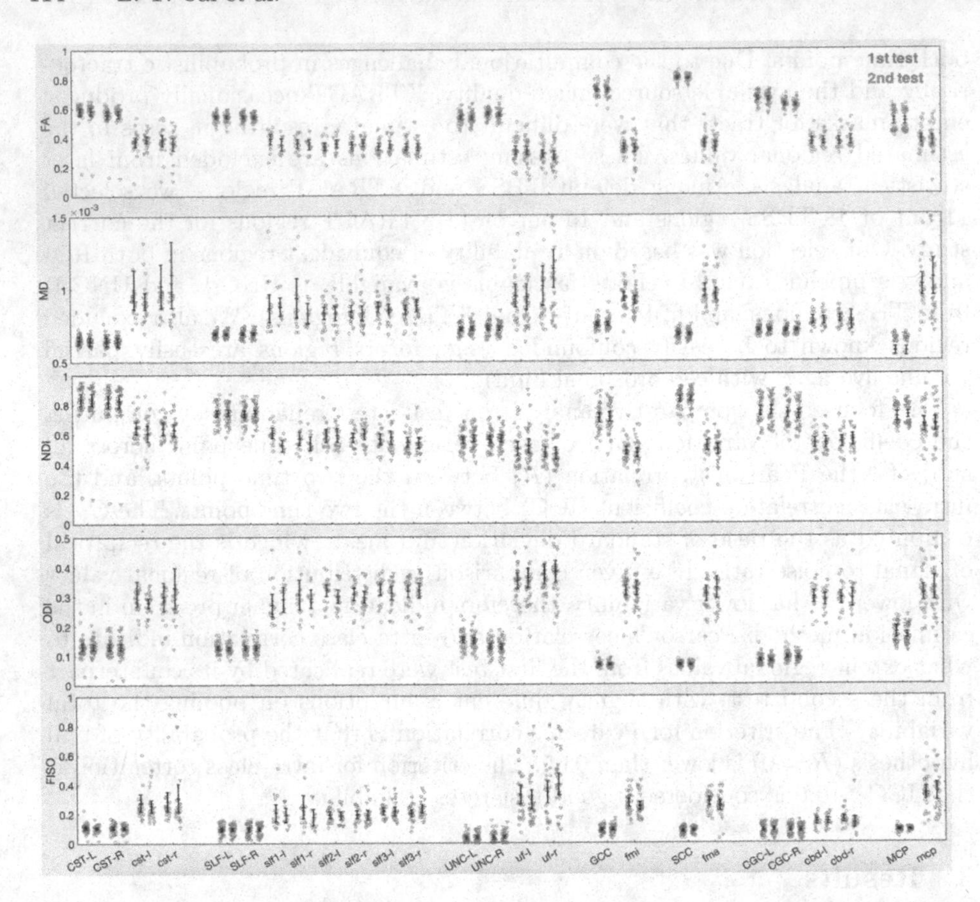

Fig. 1. Regional values from ROI analysis on DTI parameters and NODDI parameters in all subjects. For each region labeled below the bottom panel, blue circles and red triangles represent the first test and the second test, respectively. For each group of data points, the mean and the 95% confidence interval for mean estimation are indicated by black error bars.

two methods may underline any systematic patterns exhibited in Fig. 1, qualitatively summarized here: (1) regional values of FA and NDI from the TBSS pipeline are higher than counterparts from the XTRACT pipeline; (2) regional values of MD, ODI, and FISO from the TBSS pipeline are lower than counterparts from the XTRACT pipeline; (3) regional values from the XTRACT pipeline are more variable across subjects than counterparts from the TBSS pipeline; (4) regional values from the two time points in each region are visually consistent, suggesting good overall test-retest reliability of both methods; (5) most groups of data are well characterized by a unimodal distribution, although there are exceptions, for instance, the MCP might have bimodal distributions in FA and NDI values, which can be attributed to site and scanner differences.

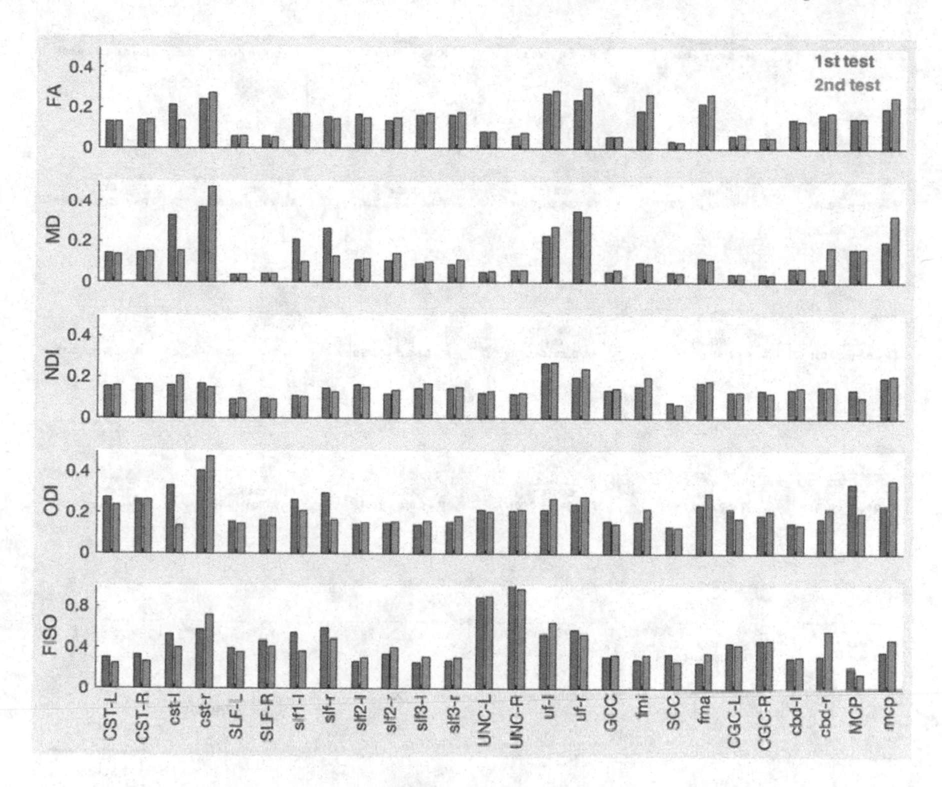

Fig. 2. Coefficient of variation of regional values.

To analyze these differences quantitatively, we computed the coefficient of variation of these regions, shown in Fig. 2. The layout of Fig. 2 is consistent with Fig. 1. Blue and red bars show C_V values from the first and second tests, respectively. From the first test, TBSS regions have significantly lower C_V in FA (t(24) = -6.86, $p < 0.001$), MD (t(24) = -3.02, $p = 0.006$), and NDI (t(24) = -2.85, $p = 0.009$), but not ODI or FISO. Consistently from the second test, these differences are also significant (t(24) = -5.68, $p < 0.001$ for FA, t(24) = -2.79, $p = 0.01$ for MD, t(24) = -3.36, $p = 0.003$ for NDI). These statistics suggest that the TBSS pipeline generally has a lower inter-subject variability therefore a higher precision in estimating the regional values from our ROI selection. In comparison, the XTRACT pipeline shows a higher inter-subject variability, but the variability is qualitatively repeatable between the two time points.

To measure the test-retest reliability quantitatively, we compute Pearson's correlation between paired regional values from the two time points. Figure 3 demonstrates this analysis by showing FA regional values from the second test plotted against those from the first test. In this figure, the 26 panels visualize the test-retest evaluation of the 26 regions, respectively, with the region name on the panel title, together with the Pearson's correlation R value and its p value against null hypothesis ($R = 0$). Regions with a star next to the title

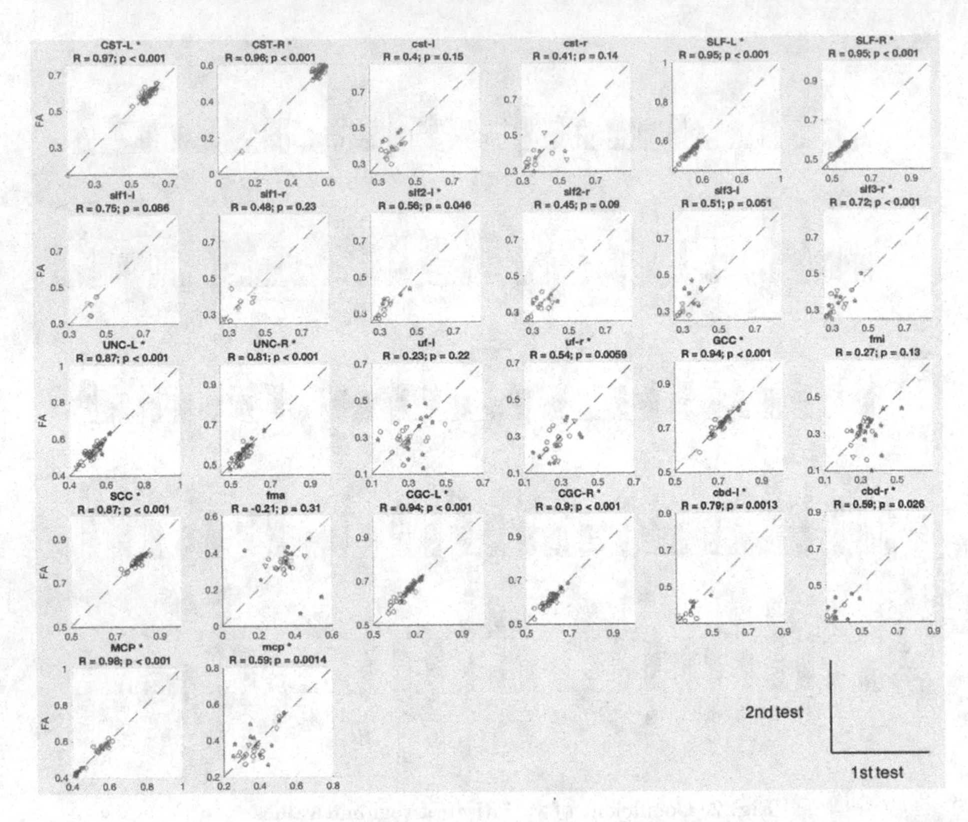

Fig. 3. Test-retest reliability of FA regional values measured by Pearson's correlation.

have a significant correlation with $p < 0.05$. The abscissa and ordinate have the same scale. The diagonal dash line on each panel shows the reference on which the values from the two time points would have had identical values. A cluster of data narrowly distributed around the reference line would have a high R, whereas a cluster of data widely scattered away from the reference line would have a low R. All regions from the TBSS pipeline but only 6 out of 15 regions from the XTRACT pipeline have significant correlations, suggesting that the TBSS pipeline has better test-retest reliability.

The advantage of the TBSS pipeline in reproducibility is true for other parameters as well. Table 2 reports all significant R values. Interestingly, NDI has shown more significant R values than any other parameters, suggesting that neurite density is a robust metric to different methods of ROI analysis.

Next, we compute the intra-class correlation coefficient (ICC) between paired regional values from the two time points. The results are shown in Fig. 4. Each bar shows the average ICC across subjects. The numbers at the very top show the number of subjects each bar value was averaged from. There was no missing estimate from the TBSS pipeline, therefore the number of subjects for ICCs of TBSS regions is always the total number of subjects (39). In contrast, the

Table 2. Test-retest reliability: Pearson's correlation between two time points.

Region	FA	MD	NDI	ODI	FISO
CST-L	$R=0.97$, $p \ll 0.001$	$R=0.95$, $p \ll 0.001$	$R=0.97$, $p \ll 0.001$	$R=0.84$, $p \ll 0.001$	$R=0.68$, $p \ll 0.001$
CST-R	$R=0.96$, $p \ll 0.001$	$R=0.92$, $p \ll 0.001$	$R=0.96$, $p \ll 0.001$	$R=0.88$, $p \ll 0.001$	$R=0.81$, $p \ll 0.001$
cst-l			$R=0.64$, $p=0.014$		
cst-r			$R=0.71$, $p=0.005$		$R=0.57$, $p=0.034$
SLF-L	$R=0.95$, $p \ll 0.001$	$R=0.90$, $p \ll 0.001$	$R=0.81$, $p \ll 0.001$	$R=0.84$, $p \ll 0.001$	$R=0.78$, $p \ll 0.001$
SLF-R	$R=0.95$, $p \ll 0.001$	$R=0.94$, $p \ll 0.001$	$R=0.85$, $p \ll 0.001$	$R=0.93$, $p \ll 0.001$	$R=0.87$, $p \ll 0.001$
slf2-l	$R=0.56$, $p=0.046$	$R=0.75$, $p=0.003$	$R=0.92$, $p \ll 0.001$		
slf2-r			$R=0.73$, $p=0.002$		
slf3-l		$R=0.54$, $p=0.037$	$R=0.67$, $p=0.006$		$R=0.60$, $p=0.018$
slf3-r	$R=0.72$, $p \ll 0.001$		$R=0.75$, $p \ll 0.001$	$R=0.66$, $p=0.003$	
UNC-L	$R=0.87$, $p \ll 0.001$	$R=0.71$, $p \ll 0.001$	$R=0.93$, $p \ll 0.001$	$R=0.77$, $p \ll 0.001$	$R=0.73$, $p \ll 0.001$
UNC-R	$R=0.81$, $p \ll 0.001$	$R=0.87$, $p \ll 0.001$	$R=0.86$, $p \ll 0.001$	$R=0.79$, $p \ll 0.001$	$R=0.75$, $p \ll 0.001$
uf-l			$R=0.52$, $p=0.004$		
uf-r	$R=0.54$, $p=0.006$		$R=0.49$, $p=0.015$	$R=0.54$, $p=0.006$	
GCC	$R=0.94$, $p \ll 0.001$	$R=0.87$, $p \ll 0.001$	$R=0.89$, $p \ll 0.001$	$R=0.60$, $p \ll 0.001$	$R=0.59$, $p \ll 0.001$
fmi		$R=0.52$, $p=0.003$	$R=0.43$, $p=0.013$		$R=0.44$, $p=0.011$
SCC	$R=0.87$, $p \ll 0.001$	$R=0.84$, $p \ll 0.001$	$R=0.92$, $p \ll 0.001$	$R=0.86$, $p \ll 0.001$	$R=0.79$, $p \ll 0.001$
fma		$R=0.61$, $p=0.001$			$R=0.43$, $p=0.028$
CGC-L	$R=0.94$, $p \ll 0.001$	$R=0.77$, $p \ll 0.001$	$R=0.96$, $p \ll 0.001$	$R=0.68$, $p \ll 0.001$	$R=0.91$, $p \ll 0.001$
CGC-R	$R=0.90$, $p \ll 0.001$	$R=0.74$, $p \ll 0.001$	$R=0.88$, $p \ll 0.001$	$R=0.63$, $p \ll 0.001$	$R=0.86$, $p \ll 0.001$
cbd-l	$R=0.79$, $p=0.001$	$R=0.76$, $p=0.003$	$R=0.88$, $p \ll 0.001$		$R=0.89$, $p \ll 0.001$
cbd-r	$R=0.59$, $p=0.025$		$R=0.91$, $p \ll 0.001$		
MCP	$R=0.98$, $p \ll 0.001$	$R=0.99$, $p \ll 0.001$	$R=0.79$, $p \ll 0.001$	$R=0.47$, $p=0.003$	$R=0.60$, $p \ll 0.001$
mcp	$R=0.59$, $p=0.001$	$R=0.61$, $p=0.001$	$R=0.73$, $p \ll 0.001$		$R=0.69$, $p \ll 0.001$

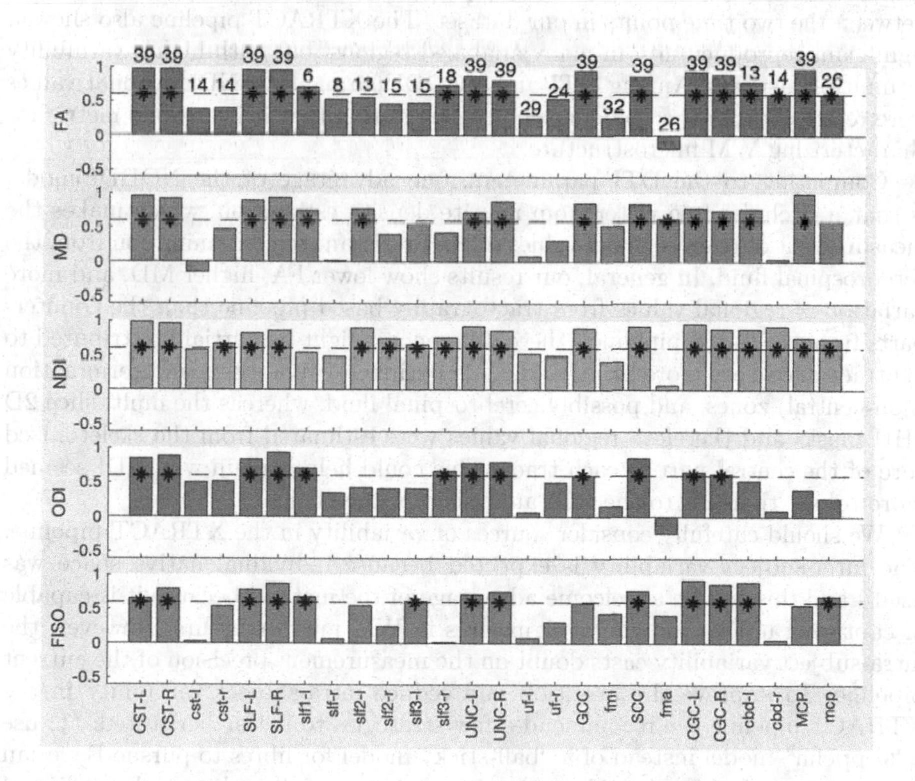

Fig. 4. Intraclass correlation coefficients between the two time points.

XTRACT pipeline has missing estimates depending on the difficulty of tractography for certain tracts. If several subjects missed either time points for a region, the number of paired data for ICC computation would be limited. With this limitation in mind, we use a criterion of ICC > 0.6 for "good reproducibility", marked as the dash line on each panel in Fig. 4. Bars meeting this criterion are marked with a black star. Across all panels, most TBSS regions (except FISO for GCC, and ODI and FISO for MCP) show good reproducibility between the two time points, suggesting trustable precision and reliability of the method. XTRACT regions do not show reproducibility as much, suggesting that the method is less stable in measuring regional values of DTI and NODDI parameters, although NDI in many XTRACT regions were robustly measured. These results are coherent with what we learned from the Pearson's correlation.

4 Discussion

In the current study, we explored the potential of performing ROI analysis of DTI and NODDI parameters in individual native diffusion space with probabilistic tractography. In the comparison between 2D tract skeleton TBSS ROI analysis in standard space and 3D whole-tract XTRACT ROI analysis in native space, the TBSS pipeline showed great precision with high test-retest reliability between the two time points in our dataset. The XTRACT pipeline also showed significant reproducibility in many analyzed regions, but with higher variability in multiple aspects. Among DTI and NODDI parameters, NDI regional values were relatively robust to pipeline choice, suggesting that it is a stable metric for characterizing WM microstructure.

Comparing to the DTI parameters, one advantage of the NODDI model is that it excludes free water from neurite density estimation, which makes the measurement of tract regional values of NDI resistant to contamination from the cerebrospinal fluid. In general, our results show lower FA, higher MD, and more variation of regional values from tractography based pipeline than the counterparts from the TBSS pipeline - these differences might be partially attributed to 3D tract masks are more likely to include peripheral (non-core) and termination (non-central) zones, and possibly cerebrospinal fluid, whereas the multi-slice 2D JHU masks and therefore regional values were estimated from the skeletonized core of the central part of each tract. This could help explain why NDI seemed more robust than FA to the ROI analysis pipeline choice.

We should carefully consider sources of variability in the XTRACT pipeline. The inter-subject variability is expected because individual native space was used - and this can be a welcome advantage of the method because it is capable of capturing unique individual signatures in WM microstructure. However, the intra-subject variability casts doubt on the measurement precision of the current pipeline. To improve the precision and reduce intra-subject variability in the XTRACT pipeline, we recommend a few strategies to be further tested: (1) use a "zeppelin" model instead of a "ball-stick" model for fibres to pursue Bayesian estimation of the crossing fibres density map, especially when high quality of dMRI data is available; (2) use FreeSurfer to reconstruct the cortical surface from the individual's T1 image volume, and use the surface as a seeding reference for tractography instead of using a standard T1 template.

Understanding the nature of probabilistic tractography as a source of analysis result variability is in agreement with the pursuit of reproducibility. Theaud et al. (2020) pointed out that fiber tracking was constructed as a stochastic process, which by nature will result in non-deterministic track segments, although one could usually make assumptions to model the distribution. In addition, software stability and random seed generation for tractography should also be carefully considered and tested in assessment of pipeline reproducibility [21].

5 Conclusion

As region-of-interest analysis of diffusion MRI metrics moves forward, comprehensive understanding of white matter microstructure demands precise and reproducible region segmentation based on advanced tractography, yet with the expectation to capture any individual specificity. In the current study, ROI analysis based on a standard atlas mapped to skeletonized tracts showed excellent precision and reproducibility; ROI analysis based on probabilistic tractography in individual diffusion space exhibited more variability both inter-subject and intra-subject, encouraging further development and exploration of the pipeline to improve precision and reliability. Neurite density index was the most robust dMRI metric to pipeline choice, however the recent literature suggests that NODDI metrics might have more cross-scanner variability than DTI metrics, especially when there are differences in the dMRI sequence such as single-band versus multi-band echoplanar imaging [7,8,17,19].

References

1. Alexander, A.L., Lee, J.E., Lazar, M., Field, A.S.: Diffusion tensor imaging of the brain. Neurotherapeutics **4**(3), 316–329 (2007)
2. Andersson, J.L., Graham, M.S., Drobnjak, I., Zhang, H., Campbell, J.: Susceptibility-induced distortion that varies due to motion: correction in diffusion MR without acquiring additional data. Neuroimage **171**, 277–295 (2018)
3. Andersson, J.L., Graham, M.S., Drobnjak, I., Zhang, H., Filippini, N., Bastiani, M.: Towards a comprehensive framework for movement and distortion correction of diffusion MR images: within volume movement. Neuroimage **152**, 450–466 (2017)
4. Andersson, J.L., Graham, M.S., Zsoldos, E., Sotiropoulos, S.N.: Incorporating outlier detection and replacement into a non-parametric framework for movement and distortion correction of diffusion MR images. Neuroimage **141**, 556–572 (2016)
5. Andersson, J.L., Skare, S., Ashburner, J.: How to correct susceptibility distortions in spin-echo echo-planar images: application to diffusion tensor imaging. Neuroimage **20**(2), 870–888 (2003)
6. Andersson, J.L., Sotiropoulos, S.N.: An integrated approach to correction for off-resonance effects and subject movement in diffusion MR imaging. Neuroimage **125**, 1063–1078 (2016)

7. Andica, C., et al.: Scan-rescan and inter-vendor reproducibility of neurite orientation dispersion and density imaging metrics. Neuroradiology **62**(4), 483–494 (2020)
8. Bouyagoub, S., Dowell, N.G., Gabel, M., Cercignani, M.: Comparing multiband and singleband EPI in NODDI at 3 T: what are the implications for reproducibility and study sample sizes? Magn. Reson. Mater. Phys. Biol. Med. **34**, 1–13 (2020)
9. Daducci, A., Canales-Rodríguez, E.J., Zhang, H., Dyrby, T.B., Alexander, D.C., Thiran, J.P.: Accelerated microstructure imaging via convex optimization (AMICO) from diffusion MRI data. NeuroImage **105**, 32–44 (2015)
10. Farquharson, S., et al.: White matter fiber tractography: why we need to move beyond DTI. J. Neurosurg. **118**(6), 1367–1377 (2013)
11. Gardner, R.C., Yaffe, K.: Epidemiology of mild traumatic brain injury and neurodegenerative disease. Mol. Cell. Neurosci. **66**, 75–80 (2015)
12. Jones, D.K., Leemans, A.: Diffusion tensor imaging. In: Modo, M.J., Bulte, J.W.M. (eds.) Magnetic Resonance Neuroimaging. Methods in Molecular Biology (Methods and Protocols), **711**, 127–144. Springer. Cham (2011). https://doi.org/10.1007/978-1-61737-992-5_6
13. Le Bihan, D., et al.: Diffusion tensor imaging: concepts and applications. J. Magn. Reson. Imaging: Official J. Int. Soc. Magn. Reson. Med. **13**(4), 534–546 (2001)
14. Lash, R.S., Bell, J.F., Reed, S.C.: Epidemiology. In: Todd, K.H., Thomas, C.R., Alagappan, K. (eds.) Oncologic Emergency Medicine, pp. 3–12. Springer, Cham (2021). https://doi.org/10.1007/978-3-030-67123-5_1
15. Lerma-Usabiaga, G., Mukherjee, P., Perry, M.L., Wandell, B.A.: Data-science ready, multisite, human diffusion MRI white-matter-tract statistics. Sci. Data **7**(1), 1–9 (2020)
16. Lerma-Usabiaga, G., Mukherjee, P., Ren, Z., Perry, M.L., Wandell, B.A.: Replication and generalization in applied neuroimaging. Neuroimage **202**, 116048 (2019)
17. Lucignani, M., Breschi, L., Espagnet, M.C.R., Longo, D., Talamanca, L.F., Placidi, E., Napolitano, A.: Reliability on multiband diffusion NODDI models: a test retest study on children and adults. NeuroImage **238**, 118234 (2021)
18. Oishi, K., et al.: Atlas-based whole brain white matter analysis using large deformation diffeomorphic metric mapping: application to normal elderly and Alzheimer's disease participants. Neuroimage **46**(2), 486–499 (2009)
19. Palacios, E.M., et al.: The evolution of white matter microstructural changes after mild traumatic brain injury: a longitudinal DTI and NODDI study. Sci. Adv. **6**(32), eaaz6892 (2020)
20. Smith, S.M., et al.: Tract-based spatial statistics: voxelwise analysis of multi-subject diffusion data. Neuroimage **31**(4), 1487–1505 (2006)
21. Theaud, G., Houde, J.C., Boré, A., Rheault, F., Morency, F., Descoteaux, M.: Tractoflow: a robust, efficient and reproducible diffusion MRI pipeline leveraging nextflow & singularity. NeuroImage **218**, 116889 (2020)
22. Warrington, S., et al.: Xtract-standardised protocols for automated tractography in the human and macaque brain. NeuroImage **217**, 116923 (2020)
23. Yue, J.K., et al.: Transforming research and clinical knowledge in traumatic brain injury pilot: multicenter implementation of the common data elements for traumatic brain injury. J. Neurotrauma **30**(22), 1831–1844 (2013)
24. Yuh, E.L., et al.: Diffusion tensor imaging for outcome prediction in mild traumatic brain injury: a track-TBI study. J. Neurotrauma **31**(17), 1457–1477 (2014)

Bundle Geodesic Convolutional Neural Network for DWI Segmentation from Single Scan Learning

Renfei Liu[✉], Francois Lauze, Kenny Erleben, and Sune Darkner

Department of Computer Science, University of Copenhagen, Copenhagen, Denmark
{renfei.liu,francois,kenny,darkner}@di.ku.dk

Abstract. We present a tissue classifier for Magnetic Resonance Diffusion Weighted Imaging (DWI) data trained from a single subject with a single b-value. The classifier is based on a Riemannian Deep Learning framework for extracting features with rotational invariance, where we extend a G-CNN learning architecture generically on a Riemannian manifold. We validate our framework using single-shell DWI data with a very limited amount of training data - only 1 scan. The proposed framework mainly consists of three layers: a lifting layer that locally represents and convolves data on tangent spaces to produce a family of functions defined on the rotation groups of the tangent spaces, i.e., a *section* of a bundle of rotational functions on the manifold, a group convolution layer that convolves this section with rotation kernels to produce a new section; and a projection layer using maximisation to collapse this local data to form new manifold based functions. We present an instantiation on the 2 dimensional sphere where the DWI orientation data is in general represented, and we use it for voxel classification. We show that this allows us to learn a classifier for cerebrospinal fluid (CSF) - subcortical - grey matter - white matter classification from only one scan.

Keywords: Single scan learning · DWI · Geodesic CNN · Classification

1 Introduction

Very little manually annotated DWI data exists, and DWI studies are, in general, small in sample size. This poses a challenge for machine learning techniques that for most parts require a significant amount of training data. However, learning from only 1 single-shell scan is possible by constructing a G-CNN architecture that takes advantage of the geometry of the data. This work focuses on building a neural network (NN) for data on manifolds with some form of orientation invariance, and here we take Diffusion Weighted Imaging as the main application. Our goal is to be able to understand spherical patterns up to rotations. There are series of proposals to generalise a $\mathbb{R}^2$ convolutional neural network to curved spaces. In general, to define convolution, the underlying space must have

S. Cetin-Karayumak et al. (Eds.): CDMRI 2021, LNCS 13006, pp. 121–132, 2021.
https://doi.org/10.1007/978-3-030-87615-9_11

a group structure or be a homogeneous space of a group. This is not always the case for curved space. But even when it is, this often imposes a certain type of filters. In our case, rotational invariance is a desirable property we want in the design. We propose a general architecture for extracting and filtering local orientation information of data defined on a manifold. The architecture allows us to learn similar orientation structures which can appear at different locations on the manifold. Reasonable manifolds have local orientation structures – rotations on tangent spaces. Our architecture lifts data to these structures and performs local filtering on them, before collapsing them back to obtained filtered features on the manifold. This provides both rotational invariance and flexibility in design, without having to resort to complex embeddings in Euclidean spaces. We provide an explicit construction of the architecture for DWI data and show very promising results for this case including single scan learning.

Related Work. The importance of the extraction of rotationally invariant features beyond Fractional Anisotropy [2] has been recognized in series of DWI works. [5] developed invariant polynomials of spherical harmonic (SH) expansion coefficients, and discussed their application in population studies. [21] proposed a related construction using eigenvalue decomposition of SH operators. [19] and [29] argued their usefulness for understanding microstructures in relation to DWI.

There is though a vast growth in literature on Deep Learning (DL) for non-flat data or more complex group actions than just translations. [17] proposed a NN on surfaces that extracts local rotationally invariant features. A non-rotationally invariant modification was proposed in [4]. On the other hand, convolution generalises to more group actions than just translation, and this has led to group-convolution neural networks for structures where these operations are supported, especially Lie groups themselves and their homogeneous spaces [1,3,9,11,15,26,27]. Global equivariance is often sought but proved complicated or even elusive in many cases when the underlying geometry is non-trivial [24]. An elementary construction on a general manifold is proposed in [20] via a fixed choice of geodesic paths used to transport filters between points on the manifold, ignoring the effects of path dependency (holonomy). Removing this dependency can be obtained by summarising local responses over local orientations, this is what is done in [17]. To explicitly deal with holonomy, [22] proposed a convolution construction on manifolds based on stochastic processes via the frame bundle, but it is at this point still very theoretical.

A few works have applied DL to DWI. [12] built multi-layer perceptrons in q-space for kurtosis and NODDI mappings. Because of the spherical structure of the DWI data and the homogeneous structure of the sphere, [6] proposed an rotation equivariant construction inspired by [8] for disease classification. [18] propose a sixth-D, 3D space and q-space NNs with roto-translation/rotation equivalence properties.

In this work, we are interested in rotationally invariant features, so we take a path closer to [17,20], but we add an extra local group convolution layer before summarising the data and eliminating path dependency. The proposed construction thus applies to oriented Riemannian manifolds, and no other structure (e.g. homogeneous or symmetric space) is used.

Organisation. We introduce the construction in the next section, first in a general setting, then in our case of interest, the sphere $\mathbb{S}^2$. We present experiments and results in Sect. 3. Discussion and conclusion are presented in Sect. 4.

2 Bundle Geodesic Convolutional Neural Network

Bekkers *et al.* [3] used the fact that $SE(2)$ acts on $\mathbb{R}^2$ to lift 2D (vector-values) images to $\mathbb{R}^2 \times \mathbb{S}^1$ via *correlation kernels.* This is not in general the case when $\mathbb{R}^2$ is replaced by an oriented Riemannian manifold $\mathcal{M}$ as there is no roto-translation group defined on a general manifold. An alternative construction is however possible by combining [3] and [17], to obtain a 3-component layer architecture: i) the **lifting layer**, ii) the **group correlation layer**, and iii) **the projection layer**. In practical applications, one or more of these multilayers can be used and a fully connected layer is built upon the last one. In this section, we focus only on the Riemannian part.

We refer the readers to [10] for the Riemannian geometric constructions. In the sequel, a base point $\boldsymbol{x}_0$ is chosen on $\mathcal{M}$. A piecewise smooth path γ joining $\boldsymbol{x}_0$ and $\boldsymbol{x} \in \mathcal{M}$ is a continuous curve that may fail to be smooth at a finite number of points. With such a curve, there is a *parallel transport* P_γ between $T_{\boldsymbol{x}_0}\mathcal{M}$ and $T_{\boldsymbol{x}}\mathcal{M}$. This is an orientation preserving isometry between tangent spaces. A tangent *kernel* at $\boldsymbol{x}_0$ is a function $\kappa : T_{\boldsymbol{x}_0}\mathcal{M} \to \mathbb{R}^N$. We assume it has a "small support". A rotational kernel at $\boldsymbol{x}_0$ is a function $K : SO(\boldsymbol{x}_0) \to \mathbb{R}^M$, where, $SO(\boldsymbol{x})$ denotes the rotation group of $T_{\boldsymbol{x}}\mathcal{M}$.

2.1 Layer Definitions

As it is usually the case that correlation replaces convolution in convolutional neural networks (CNN). The first two layers will be defined via correlation.

Lifting Layer. The correlation $f \tilde{\star}_\gamma \kappa$ of $f \in L^2(\mathcal{M}, \mathbb{R}^N)$ is defined as the function on $SO(\boldsymbol{x})$

$$f \tilde{\star}_\gamma \kappa(S) = \sum_{i=1}^{N} \int_{T_{\boldsymbol{x}}\mathcal{M}} \kappa_i(P_\gamma^{-1} S^{-1} v) f_i(\mathrm{Exp}_{\boldsymbol{x}}(v))\, dv \tag{1}$$

We assume that $\kappa \circ P_\gamma^{-1}$, the support of κ, is sufficiently small so that the exponential map is injective. For any other path δ between $\boldsymbol{x}_0$ and $\boldsymbol{x}$, it is easy to show that there exists a rotation $R \in SO(T_{\boldsymbol{x}}\mathcal{M})$ that only depends on P_γ and P_δ with $f \tilde{\star}_\delta \kappa(S) = f \tilde{\star}_\gamma \kappa(RS)$. For any point $\boldsymbol{x}$ and a path $\gamma_{\boldsymbol{x}}$ between $\boldsymbol{x}_0$ and $\boldsymbol{x}$, this filters/lifts f to functions $F_{\boldsymbol{x}} : SO(\boldsymbol{x}) \to \mathbb{R}$. Using an input

$f^{(\ell-1)}: \mathcal{M} \to \mathbb{R}^{N_{\ell-1}}$ and N_ℓ $\boldsymbol{x}_0$-kernels $\boldsymbol{\kappa}^{(\ell)} = \left(\kappa_1^{(\ell)}, \ldots, \kappa_{N_\ell}^{(\ell)}\right)$, $\kappa_i^{(\ell)}$ $\mathbb{R}^{N_{\ell-1}}$-valued at layer $\ell - 1$,

$$\forall \boldsymbol{x} \in \mathcal{M}, \quad F_{\boldsymbol{x}}^{(\ell)} = \left(f^{(\ell-1)} \tilde{\star}_{\gamma_{\boldsymbol{x}}} \kappa^1, \ldots, f^{(\ell-1)} \tilde{\star}_{\gamma_{\boldsymbol{x}}} \kappa_{N_\ell}\right) \tag{2}$$

The output $F^{(\ell)}$ is not a function defined on $\mathcal{M}$, but a *section*, in general non smooth, of the *function bundle* $\mathbb{L}^2\left(SO(\mathcal{M}), \mathbb{R}^{N_\ell}\right) = \sqcup_{\boldsymbol{x}} L^2(SO(\boldsymbol{x}), \mathbb{R}^{N_\ell})$.

Group Correlation Layer. if F is a function $SO(\boldsymbol{x}) \to \mathbb{R}^M$, we define $F \star_\gamma K$ as the *classical* group correlation

$$F \star_\gamma K(S) = \sum_{i=1}^{M} \int_{SO(\boldsymbol{x})} F_i(U) K_i(P_\gamma^{-1} S^{-1} U P_\gamma)\, dU. \tag{3}$$

This construction provides a new family of functions $\bar{F}_{\boldsymbol{x}} : SO(\boldsymbol{x}) \to \mathbb{R}$. Differing from [3], translations are in general not defined in $\mathcal{M}$ and rotations are only local. If $F_\gamma = (f_i \tilde{\star}_\gamma \kappa_i)_{i=1}^{M}$ and $F_\delta = (f_i \tilde{\star}_\delta \kappa_i)_{i=1}^{M}$ then it can be easily shown using the bi-invariance of the Haar measure on $SO(n)$ that $\varphi(F_\delta) \star_\delta K(S) = \varphi(F_\gamma) \star_\gamma K(SR)$ where R depends only on paths γ and δ, and φ is any real function (typically a rectified linear unit (ReLU)). With input $F^{(\ell-1)} \in \mathbb{L}^2\left(SO(\mathcal{M}), \mathbb{R}^{N_{\ell-1}}\right)$ with $N_{\ell-1}$ channels at layer $\ell - 1$ and $\boldsymbol{x}_0$-rotation kernels $\boldsymbol{K}^{(\ell)} = \left(K_1^{(\ell)}, \ldots, K_{N_\ell}^{(\ell)}\right)$, each with $N_{\ell-1}$ channels, one obtains $F^{(\ell)} \in \mathbb{L}^2(SO(\mathcal{M}), \mathbb{R}^{N_\ell})$ as

$$F_{\boldsymbol{x}}^{(\ell)} = \left(F^{(\ell-1)} \star_{\gamma_{\boldsymbol{x}}} K_1^{(\ell)}, \ldots, F^{(\ell-1)} \star_{\gamma_{\boldsymbol{x}}} K_{N_\ell}^{(\ell)}\right) \tag{4}$$

Projection Layer. A family $F^{(\ell-1)} \in \mathbb{L}^2\left(SO(\mathcal{M}), \mathbb{R}^{N_{(\ell-1)}}\right)$ is projected to a function $f : \mathcal{M} \to \mathbb{R}^{N_{(\ell-1)}}$ as

$$f_i^{(\ell)}(x) = \max_{S \in SO(\boldsymbol{x})} F_{i\boldsymbol{x}}^{(\ell-1)}(S), \quad i = 1 \ldots N_{\ell-1} \tag{5}$$

This removes the path dependency thanks to the change of path property which was described above. See Fig. 1a for illustration.

Biases are added per kernel. Nonlinear transformations of ReLU type are applied after each of these layers. Note that without them, a lifting followed by group correlation would actually factor in a new lifting transformation.

2.2 Discretisation and Implementation in the Case $\mathcal{M} = \mathbb{S}^2$

In this work, the manifold of interest is $\mathbb{S}^2$. Spherical functions $f : \mathbb{S}^2 \to \mathbb{R}^N$ are typically given at a number of points and interpolated using a Watson kernel [13], which also serves as our choice. We use a very simple discretisation of $\mathbb{S}^2$ via the vertices of a regular icosahedron. Tangent kernels are defined over these vertices, sampled along with the rays of a polar coordinate system respecting the vertices of the icosahedron. This is illustrated in Fig. 1b.

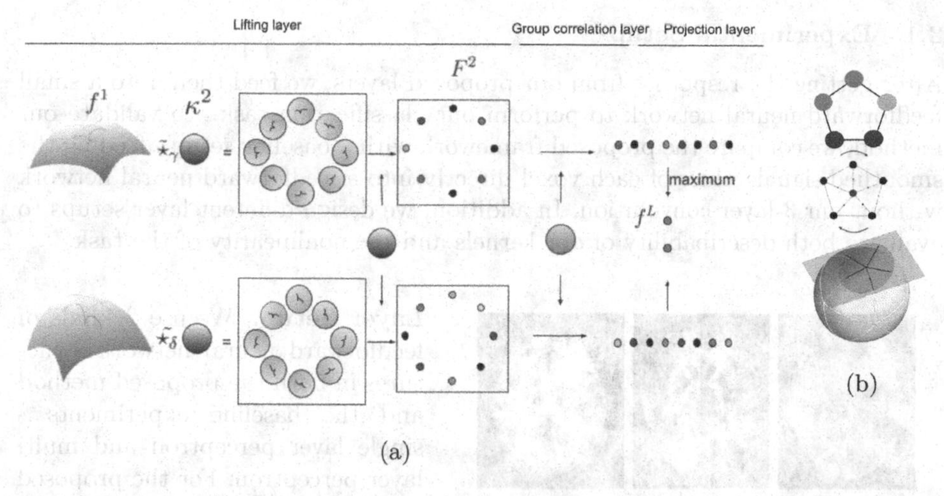

Fig. 1. In Fig. a, the top row shows the lifting kernel $\kappa^{(2)}$ applied at a point on the manifold, resulting in an image $F^{(2)}$ defined on $SO(2)$ as in Eq. 1. The function is first mapped onto the tangent space of the point of interest via the exponential map, and $\kappa^{(2)}$ is convolved with the mapped function to get F^2. In the figure we rotate the tangent space instead of the kernel as Eq. 1 for convenience in illustration, but they are equivalent constructions. We get rotationally invariant responses from the projection layer. The bottom row shows the same process but with a different kernel parallel transport, illustrating that the responses of the convolutional layers are simply rotated. In Fig. b, the bottom row shows $\mathbb{S}^2$ with a regular icosahedric tessellation and a tangent plane at one of the vertices and 5 sampled directions. The disk represents the kernel support. The middle row shows the actual discrete kernel used, with the $2\pi/5$ rotations and the top row is represents the lifted function on the discrete rotation group.

3 Experiments and Results

We evaluate our method on DWI data from the human connectome project [25]. We train a network using our framework on individual voxels containing signals on $\mathbb{S}^2$. Our goal is a voxel-wise classification of 4 regions of the brain - cerebrospinal fluid (CSF), subcortical, white matter, and grey matter regions.

We used the pre-processed DWI data [23] and normalised each DWI scan for the b-1000, b-2000, and b-3000 images respectively with the voxel-wise average of the b_0. The labels provided with the T1-image were transformed to the DWI using nearest neighbour interpolation (Fig. 2). Since the 4 brain regions we are classifying have imbalanced numbers of voxels, we use Focal Loss [16] to counter the class imbalance of the dataset.

3.1 Experimental Setup

After getting the responses from our proposed layers, we feed them into a small feedforward neural network to perform our classification task. To validate our method, we compare the proposed framework with a baseline setup - feeding the smoothed signal values of each voxel directly into a feedforward neural network without our 3-layer convolution. In addition, we design different layer setups to evaluate both describability of our kernels and the nonlinearity of the task.

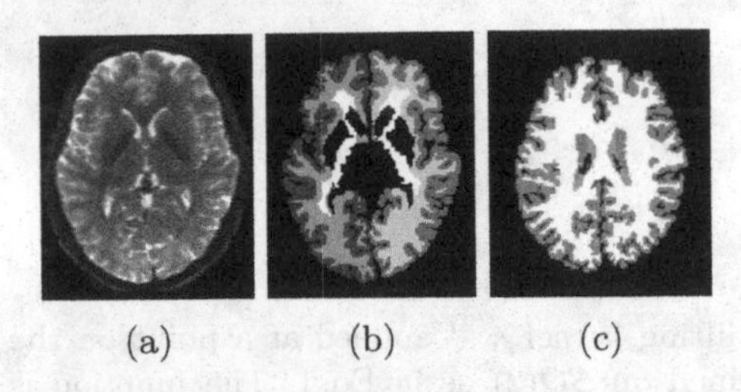

Fig. 2. (a)–(c): original diffusion data, the ground-truth segmentation, and the processed ground-truth that we are going to learn from. The label colors for CSF, subcortical, white matter and grey matter are red, blue, white and grey respectively. The figures are only for illustrations of the data, they are not necessarily from the same slice of the same scan. (Color figue online)

Layer Setup. We use 2 kinds of feedforward neural network structures in both the proposed method and the baseline experiments - single layer perceptron and multi layer perceptron. For the proposed method, connecting the responses from our 3-layer structure to a single layer perceptron without hidden layers tests the describability of our kernels while connecting the responses to a multi layer perceptron explores the full capacity of the method for the task. For baseline, feeding the smoothed signals directly to a single layer perceptron simply showcases the nonlinearity of the task, and the multi layer perceptron setup, with nonlinearity added to the model, provides a generic comparison to the proposed method as a nonlinear model without dealing with the geometric encoding of the data. The network structures for both experiment setups are given in Table 1. For full models, each linear layer is followed by a ReLU activation and batch normalisation. For all models, the output at the last layer is followed by a softmax function to generate a probability map for the 4 classes.

In order to keep the model simple, we use the icosahedron structure as kernel locations with relatively low orientation resolution of the kernels - 5 rays per kernel, and 2 sample points per ray. The radius of the kernels should guarantee that the kernel coverage of 2 adjacent icosahedron vertices will overlap with each other, therefore we choose 0.6 as our radius. We use 1 kernel for the lifting layer, and 5 kernels for the group convolution layer to select 5 most essential structures of the signals.

3.2 Results

We use **1** scan for training, **1** scan for validation, and **50** scans for testing, all of which are with single-shell setup. We have observed that the overall validation loss and accuracy for all experiments converge after a few epochs. However, for

Table 1. Illustration of the layer setups and degrees of freedom for our experiments.

Experiment	Layer setup	
	Reduced model	Full model
Baseline	(90, 4), DOF: 364	(90, 50, 30, 4), DOF: 6364
Proposed method	(60, 4), DOF: 286	(60, 30, 4), DOF: 2056

the proposed method, for inner-class accuracies of difficult minority classes such as the subcortical region, the accuracy rises gradually towards convergence while not affecting the overall accuracy at all. Therefore, we use the convergence of subcortical region classification accuracy as a stopping criterion, which occurred after around 30 epochs. We train each network presented in Table 1 for 25 epochs with batch size 100 on an Ubuntu 20.04.2 LTS machine with an Intel Xeon(R) Silver 4210 CPU @ 2.20 GHz - 40 processor and a GEFORCE RTX 3090 graphics card. Our framework is implemented in Python 3.6 and Pytorch 1.7. For the proposed method, it takes around 4 min for an epoch and 1261 MiB GPU memories for both experiment setups. For baseline experiments, it takes around 2 min and 3 min for an epoch, and 1261 Mib and 1279 Mib for the reduced and full model experiments respectively. We use $\kappa = 10$ for the interpolation using Watson kernel, and $\gamma = 2, \alpha = (0.15, 0.15, 0.35, 0.35)$ for focal loss, where the α weights correspond to CSF, subcortical regions, white matter, and grey matter respectively. Overall and inner-class accuracies and Dice scores are shown in Table 2.

Firstly, across different b values, we observe that with increased b, over all experiments, it becomes harder to recognise CSF. Secondly, across different experiment setups, we see from the single layer perceptron baseline experiment that the model performance improves with increasing b, yet the most difficult region to recognise - subcortical - was almost ignored by the model for all b. Higher b provides more distinguishable signals for the majority classes, which contributes to the overall accuracy. Counterintuitively, the addition of nonlinearity to the baseline experiment - using multi layer perceptron - even worsens the performance. Adding nonlinearity does make the recognition of subcortical region more robust, but it is at a high cost of the misclassification of other classes, which is also why the Dice score for subcortical is still low while the accuracy is high for this experiment across all b values (see Fig. 3b). On the other hand, our proposed method shows robust performance generalising to the test set with far fewer degrees of freedom, and is stable across different b. Additionally, the full model of the proposed method - connecting our convolution layers to a multi layer perceptron - does a better job in recognising the subcortical region than its reduced model counterpart without causing much misclassification of other classes as in the baseline experiments. This shows that recognising a difficult class requires geometric structure as well as higher degrees of freedom of the model. See Fig. 3 for distributions of accuracies and Dice scores of the 4 classes

Table 2. Results of both baseline and proposed method.

Results (Proposed/Baseline)	Layer setup	
	Reduced model	Full model
b = 1000		
Overall accuracy	0.78/0.598	**0.798**/0.533
CSF accuracy, Dice	0.782/0.717, 0.782/0.768	0.769/**0.827**, **0.783**/0.728
Subcortical accuracy, Dice	0.18/0.014, 0.21/0.026	0.353/**0.689**, **0.348**/0.171
White matter accuracy, Dice	0.706/0.639, 0.777/0.601	**0.767**/0.632, **0.805**/0.709
Grey matter accuracy, Dice	**0.914**/0.628, 0.833/0.627	0.877/0.419, **0.844**/0.573
b = 2000		
Overall accuracy	**0.791**/0.71	0.779/0.562
CSF accuracy, Dice	0.706/0.68, **0.725**/0.709	0.676/**0.767**, 0.699/0.634
Subcortical accuracy, Dice	0.027/0.034, 0.048/0.052	0.433/**0.592**, **0.353**/0.163
White matter accuracy, Dice	0.778/0.628, **0.802**/0.684	0.734/**0.779**, 0.791/0.796
Grey matter accuracy, Dice	**0.895**/0.861, 0.83/0.773	0.862/0.364, **0.831**/0.518
b = 3000		
Overall accuracy	**0.78**/0.775	0.774/0.698
CSF accuracy, Dice	**0.627**/0.46, **0.62**/0.554	0.314/0.262, 0.414/0.301
Subcortical accuracy, Dice	0.271/0.002, 0.28/0.004	0.363/**0.407**, **0.33**/0.21
White matter accuracy, Dice	0.727/0.779, 0.791/0.779	0.717/**0.822**, 0.787/**0.805**
Grey matter accuracy, Dice	**0.891**/0.873, **0.831**/0.822	0.887/0.638, 0.827/0.73

across 50 test scans. We show statistics in Fig. 3 for *b*-1000 scans, which are the most common single-shell data.

Another fact that is worth mentioning is that for *b*-3000, the validation accuracy for CSF fluctuates drastically for all experiments except for the reduced baseline model. This, in our opinion, is due to the fact that with much higher noise in the data, it becomes harder to recognise the diffusion in CSF in general, and stepping into a local minimum in the nonlinear models that disregards CSF will not cost much the overall loss since it is a minority class.

Moreover, we have also trained both the proposed method and the baseline with 10 scans for *b*-1000 to test how much the size of the dataset is influencing the results. It has mildly improved the classification accuracies of our method to around 0.8 and 0.81 for the reduced and full model respectively but shows no significant improvement for the baseline experiments. Therefore, we can conclude that our method can capture the most significant features with a very limited amount of data while the standard neural networks suffer from capturing geometric features even when the dataset is significantly increased.

See Fig. 4 for examples of predictions from both proposed and baseline models, with *b*-1000.

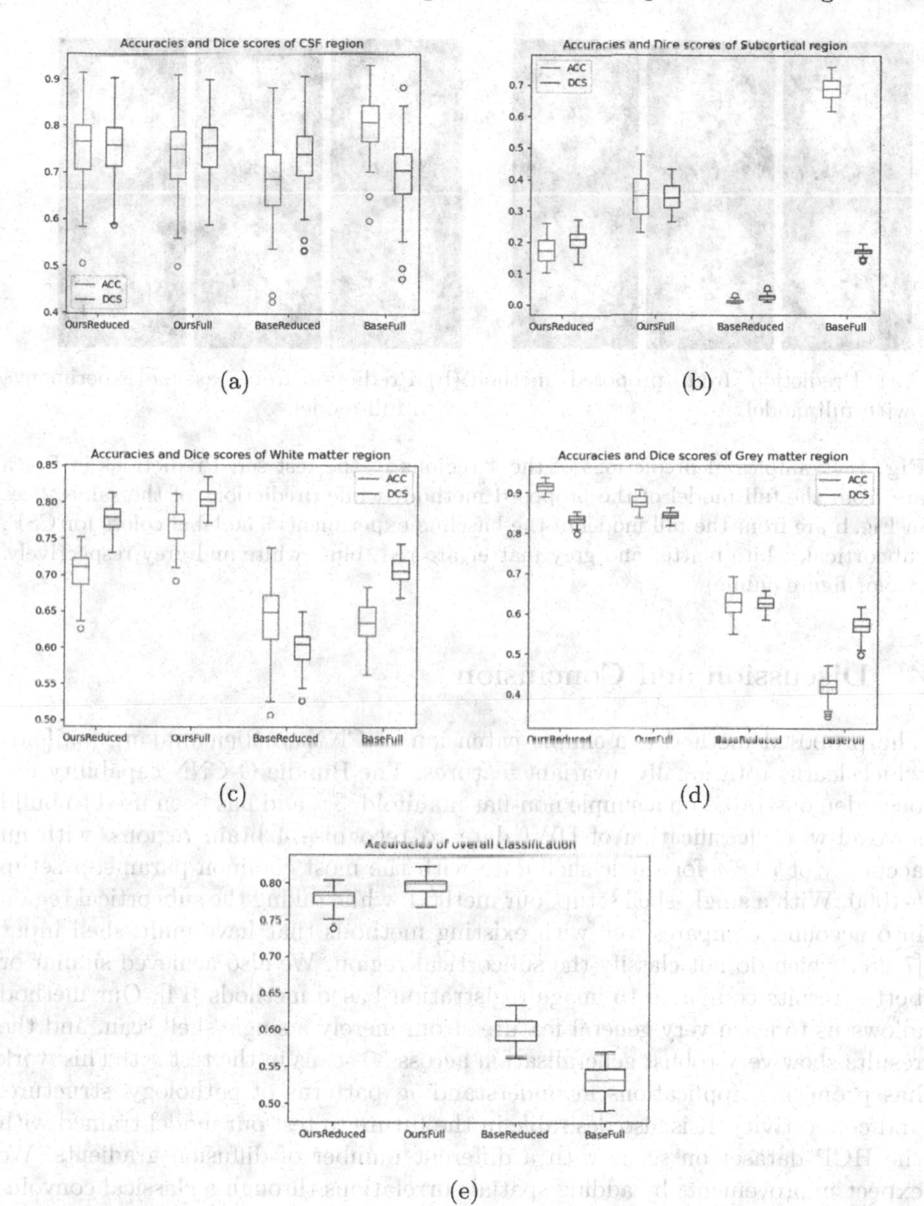

Fig. 3. Statistics of accuracy and Dice score across individual scans of the 4 regions in the test set for b-1000.

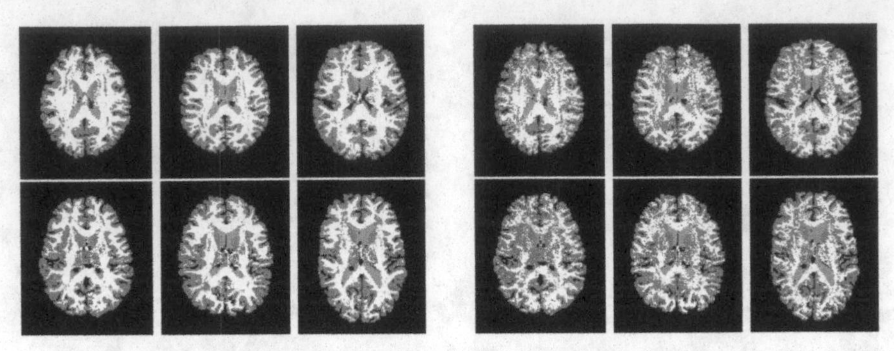

(a) Prediction from proposed method with full model.

(b) Prediction from baseline experiment with full model.

Fig. 4. Examples of predictions of the 4 regions in the test set. Predictions in Fig. a are from the full model of the proposed method, while predictions of the same slices in Fig. b are from the full model of the baseline experiment. The label colors for CSF, subcortical, white matter and grey matter are red, blue, white and grey respectively. (Color figure online)

4 Discussion and Conclusion

The proposed method is a simple extension of CNN to Riemannian Manifolds which learns rotationally invariant features. The Bundle G-CNN capability has been demonstrated on a simple non-flat manifold, $\mathbb{S}^2$, and has been used to build a voxel-wise classification of DWI data to recognise 4 brain regions, with an accuracy of 79.8% for single-shell data with the most common parameter setup b-1000. With a single-shell setup, our method, while taking the subcortical region into account, compares well with existing methods that have multi-shell input [7,28], which do not classify the subcortical region. We also achieved similar or better results compared to image registration based methods [14]. Our method allows us to learn very general features from merely a single-shell scan, and the results show very robust generalisation across 50 scans in the test set. This work has promising applications in understanding patterns of pathology, structure, and connectivity. It is also desirable in the future to test our model trained with the HCP dataset on scans with a different number of diffusion gradients. We expect improvements by adding spatial correlations through a classical convolutional layer, and the correlation of our model to fractional anisotropy (FA) and NODDI is worth investigating as well. Additionally, we have so far only tested it on $\mathbb{S}^2$, however, an extension to other surfaces appears feasible, though the choice of a discrete representation is important. An extension to dimension 3 will require efficient $SO(3)$ convolutions, using, for instance, spectral theory for compact Lie groups.

Acknowledgements

This project has received funding from the European Union's Horizon 2020 research and innovation programme under the Marie Sklodowska-Curie grant agreement No. 801199. This paper only contains the author's views and the Research Executive Agency and the Commission are not responsible for any use that may be made of the information it contains.

Data were provided [in part] by the Human Connectome Project, WU-Minn Consortium (Principal Investigators: David Van Essen and Kamil Ugurbil; 1U54MH091657) funded by the 16 NIH Institutes and Centers that support the NIH Blueprint for Neuroscience Research; and by the McDonnell Center for Systems Neuroscience at Washington University.

This project is also partially funded by 3Shape A/S.

References

1. Andrearczyk, V., Fageot, J., Depeursinge, A.: Local rotation invariance in 3D CNNs. Med. Image Anal. **65** (2020)
2. Basser, P., Mattiello, J., LeBihan, D.: MR diffusion tensor spectroscopy and imaging. Biophys. J. **66**(1), 259–267 (1994)
3. Bekkers, E.J., Lafarge, M.W., Veta, M., Eppenhof, K.A.J., Pluim, J.P.W., Duits, R.: Roto-translation covariant convolutional networks for medical image analysis. In: Frangi, A.F., Schnabel, J.A., Davatzikos, C., Alberola-López, C., Fichtinger, G. (eds.) MICCAI 2018. LNCS, vol. 11070, pp. 440–448. Springer, Cham (2018). https://doi.org/10.1007/978-3-030-00928-1_50
4. Boscaini, D., Masci, J., Rodolà, E., Bronstein, M.: Learning shape correspondence with anisotropic convolutional neural networks. In: Lee, D., Sugiyama, M., Luxburg, U., Guyon, I., Garnett, R. (eds.) Advances in Neural Information Processing Systems, vol. 29 (2016)
5. Caruyer, E., Verma, R.: On facilitating the use of HARDI in population studies by creating rotation-invariant markers. Med. Image Anal. **20**(1), 87–96 (2015)
6. Chakraborty, R., Banerjee, M., Vemuri, B.: A CNN for homogeneous Riemannian manifolds with application to NeuroImaging (2018)
7. Cheng, H., Newman, S., M. Afzali, S.S.F., Garyfallidis, E.: Segmentation of the brain using direction-averaged signal of DWI images. Magn. Resonan. Imaging **69**, 1–7 (2020)
8. Cohen, T., Geiger, M., Köhler, J., Welling, M.: Spherical CNNs. In: Proceedings of ICLR (2018)
9. Cohen, T., Welling, M.: Group equivariant convolutional neural networks. In: International Conference on Machine Learning, pp. 2990–2999 (2016)
10. Petersen, P.: Riemannian Geometry. GTM, vol. 171. Springer, Cham (2016). https://doi.org/10.1007/978-3-319-26654-1_9
11. Gens, R., Domingos, P.: Deep symmetry networks. In: NIPS, pp. 2537–2545 (2014)
12. Golkov, V., et al.: *q*-Space deep learning: twelve-fold shorter and model-free diffusion MRI scans. IEEE Trans. Med. Imaging **35**(5), 1344–1351 (2016)
13. Jupp, P.E., Mardia, K.V.: A unified view of the theory of directional statistics, 1975–1988. Int. Stat. Rev./Revue Internationale de Statistique **57**(3), 261–294 (1989)

14. Klein, A., et al.: Evaluation of 14 nonlinear deformation algorithms applied to human brain MRI registration. Neuroimage **46**(3), 786–802 (2009)
15. Kondor, R., Trivedi, S.: On the generalization of equivariance and convolution in neural networks to the action of compact groups. In: Proceedings of ICML, pp. 2747–2755 (2018)
16. Lin, T.Y., Goyal, P., Girshick, R., He, K., Dollár, P.: Focal loss for dense object detection (2018)
17. Masci, J., Boscaini, D., Bronstein, M., Vandergheynst, P.: Geodesic convolutional neural networks on Riemannian manifolds. In: Proceeding of 3DRR (2015)
18. Müller, P., Golkov, V., Tomassini, V., Cremers, D.: Rotation-equivariant deep learning for diffusion MRI (2021)
19. Novikov, D., Veraart, J., Jelescu, I., Fieremans, E.: Rotationally-invariant mapping of scalar and orientational metrics of neuronal microstructure with diffusion MRI. NeuroImage **174**, 518–538 (2018)
20. Schonsheck, S.C., Dong, B., Lai, R.: Parallel transport convolution: a new tool for convolutional neural networks on manifolds (2018)
21. Schwab, E., Cetingül, H.E., Asfari, B., Vidal, E.: Rotational invariant features for HARDI. In: Proceedings of IPMI (2013)
22. Sommer, S., Bronstein, A.: Horizontal flows and manifold stochastics in geometric deep learning. IEEE Trans. PAMI (2020)
23. Sotiropoulos, S., et al.: Effects of image reconstruction on fiber orientation mapping from multichannel diffusion MRI: reducing the noise floor using SENSE. Magn. Resonan. Med. **70**(6), 1682–1689 (2013)
24. Cohen, T.S., Weiler, M., Kicanaoglu, B., Welling, M.: Gauge equivariant convolutional networks and the icosahedral CNN. In: Proceedings of ICML, pp. 1321–1330 (2019)
25. Van Essen, D.C., Smith, S.M., Barch, D.M., Behrens, T.E., Yacoub, E., Ugurbil, K.: The WU-Minn human connectome project: an overview. NeuroImage **80**, 62–79 (2013)
26. Weiler, M., Geiger, M., Welling, M., Boomsma, W., Cohen, T.: 3D steerable CNNs: learning rotationally equivariant features in volumetric data. In: Proceedings of NIPS (2018)
27. Worrall, D., Garbin, S., Turmukhambetov, D., Brostow, G.: Harmonic networks: deep translation and rotation equivariance (2017)
28. Yap, P.T., Zhang, Y., Shen, D.: Brain tissue segmentation based on diffusion MRI using $\ell 0$ sparse-group representation classification. In: Navab, N., Hornegger, J., Wells, W., Frangi, A. (eds.) MICCAI 2015. LNCS, vol. 9351, pp. 132–139. Springer, Cham (2015). https://doi.org/10.1007/978-3-319-24574-4_16
29. Zucchelli, M., Deslauriers-Gauthier, S., Deriche, R.: A computational framework for generating rotation invariant features and its application in diffusion MRI. Med. Image Anal. **60** (2020)

Lesion Normalization and Supervised Learning in Post-traumatic Seizure Classification with Diffusion MRI

Md Navid Akbar[1(✉)], Sebastian Ruf[1], Marianna La Rocca[2], Rachael Garner[2], Giuseppe Barisano[3], Ruskin Cua[4], Paul Vespa[5], Deniz Erdoğmuş[1], and Dominique Duncan[2]

[1] Department of Electrical and Computer Engineering, College of Engineering, Northeastern University, Boston, MA 02115, USA
makbar@ece.neu.edu

[2] USC Stevens Neuroimaging and Informatics Institute, Keck School of Medicine, University of Southern California, Los Angeles, CA 90033, USA

[3] Zilkha Neurogenetic Institute, Keck School of Medicine, University of Southern California, Los Angeles, CA 90033, USA

[4] Department of Radiology, Keck School of Medicine, University of Southern California, Los Angeles, CA 90033, USA

[5] David Geffen School of Medicine, University of California Los Angeles, Los Angeles, CA 90095, USA

Abstract. Traumatic brain injury (TBI) is a serious condition, potentially causing seizures and other lifelong disabilities. Patients who experience at least one seizure one week after TBI (late seizure) are at high risk for lifelong complications of TBI, such as post-traumatic epilepsy (PTE). Identifying which TBI patients are at risk of developing seizures remains a challenge. Although magnetic resonance imaging (MRI) methods that probe structural and functional alterations after TBI are promising for biomarker detection, physical deformations following moderate-severe TBI present problems for standard processing of neuroimaging data, complicating the search for biomarkers. In this work, we consider a prediction task to identify which TBI patients will develop late seizures, using fractional anisotropy (FA) features from white matter tracts in diffusion-weighted MRI (dMRI). To understand how best to account for brain lesions and deformations, four preprocessing strategies are applied to dMRI, including the novel application of a lesion normalization technique to dMRI. The pipeline involving the lesion normalization technique provides the best prediction performance, with a mean accuracy of 0.819 and a mean area under the curve of 0.785. Finally, following statistical analyses of selected features, we recommend the dMRI alterations of a certain white matter tract as a potential biomarker.

This project was funded by the National Institute of Neurological Disorders and Stroke (NINDS) of the National Institutes of Health (NIH) under award number R01NS111744.

S. Cetin-Karayumak et al. (Eds.): CDMRI 2021, LNCS 13006, pp. 133–143, 2021.
https://doi.org/10.1007/978-3-030-87615-9_12

Keywords: Post-traumatic epilepsy · Diffusion MRI · Lesion normalization · Feature selection · Classification · Biomarker

1 Introduction

Traumatic brain injury (TBI), a condition in which physical injury to the brain causes temporary or permanent impairment to brain function [18], can have far-reaching consequences [12]. A particular concern is the development of seizures in response to the injury, which can indicate a diagnosis of post-traumatic epilepsy (PTE)[9]. This work studies the development of seizures with subjects from the Epilepsy Bioinformatics Study for Antiepileptogenic Therapy (EpiBioS4Rx) which aims to identify accurate biomarkers of epileptogenesis and subsequently design and perform preclinical trials of antiepileptogenic therapies. Reliable biomarkers could then serve as surrogate endpoints for future clinical trials of antiepileptogenic therapies in humans.

Magnetic resonance imaging (MRI) methods that probe structural and functional alterations after TBI are promising for biomarker detection [14]. Diffusion-weighted MRI (dMRI) is one method that researchers have used to identify changes in white matter (WM) connectivity that relate to seizure development [10]. Many dMRI studies are based on estimating quantitative indices such as fractional anisotropy (FA), mean diffusivity (MD), and apparent diffusion coefficient (ADC), which characterize water diffusion in WM tracts [11,20]. Of these indices, only FA is found to be a promising biomarker for epilepsy and TBI, whereas MD and ADC are not [11,20]. As such, we focus our exploration in this paper on FA features.

FA has been used to distinguish and classify subtypes of temporal lobe epilepsy (TLE) in a region of interest (ROI) based analysis with perfect accuracy, using logistic regression [17]. Other ROI based studies have modeled both functional and structural connectivity derived from functional MRI (fMRI) and dMRI, respectively, to predict seizure outcomes in postsurgery TLE patients with 100% accuracy, using principal component analysis (PCA), correlation, and Euclidean distance [16]. FA features extracted from voxel-wise analysis have also been applied for the prediction of consciousness recovery with an 86% accuracy, using the machine learning (ML) technique of linear discriminant analysis (LDA) [20]. Even though a perfect classification is reported in both [16,17], the subjects did not have any lesions and the imaging data in each was acquired in a single site, with the same scanner. FA features obtained from tract based spatial statistic (TBSS) have also been used to train support vector machine (SVM) ML classifiers, to characterize seizure development with 85.7% accuracy [3].

To the best of our knowledge, none of these works compensate for brain lesions and deformations that often occur with moderate-severe TBI, and that are not well corrected by the standard nonlinear registration [20].

In this work, we explore methods to address TBI induced deformation through dMRI preprocessing, including the application of lesion normalization techniques to dMRI scans obtained from multiple sites, for patients with complete follow-up data. Here we perform cost function masking [4], a technique

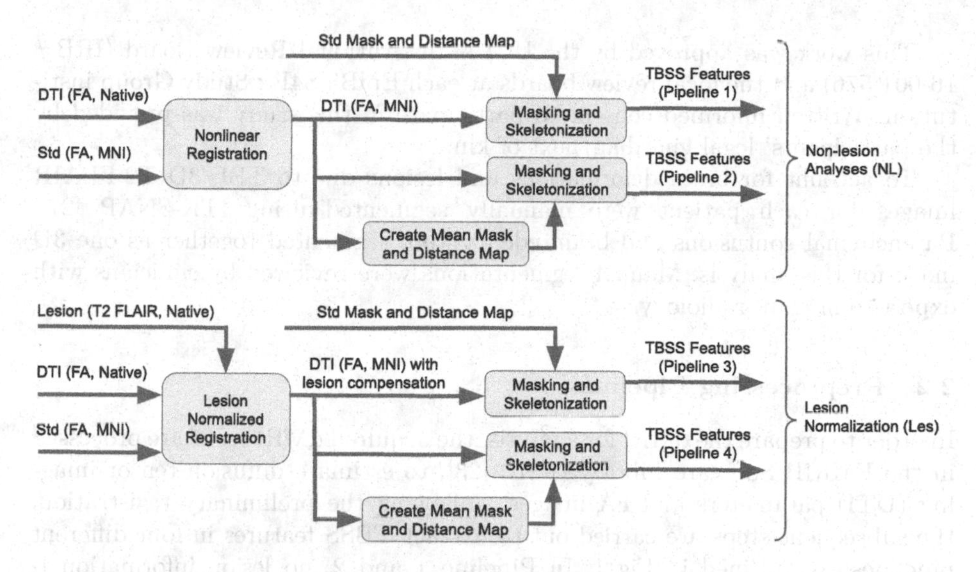

Fig. 1. The four different TBSS preprocessing pipelines.

normally used for fMRI analysis, when normalizing the dMRI into Montreal Neurological Institute (MNI) space. Once the dMRIs are in MNI space, we systematically extract FA features of different WM tracts from four parallel preprocessing pipelines, two of which utilize the proposed lesion normalization technique. We then test ML models encompassing different feature selection methods and binary classifiers to comprehensively evaluate the predictive performance of each pipeline, and statistically test the discriminative power of the features selected by the best ML model of each pipeline. Finally, we recommend the combination of the preprocessing pipeline and the ML model that yields the highest mean accuracy and mean area under curve (AUC) as a promising tool for the early prediction of seizure occurrence in TBI patients, while recommending any feature that obtains statistical significance as a potential biomarker.

2 Methods

2.1 DMRI Acquisition and Lesion Segmentation

According to the EpiBioS4Rx protocol [2], moderate-severe TBI patients with evidence of contusion were eligible for enrollment. A total of 22 patients (18 male, 4 female; average age = 45.0, SD = 20.9) were chosen for this work: 13 (63%) experienced at least one seizure more than 7 days after injury, whereas the remaining 9 were seizure-free. Mean Glasgow Coma Scale at emergency department arrival was 9.4 (SD = 4.1) Multimodal imaging was acquired, on average, 15 days post-injury (SD = 9.1). The acquired imaging included, but not limited to, different MRI sequences such as T1 MRI, dMRI, T2, and T2-weighted fluid attenuated inversion recovery (FLAIR).

This work was approved by the UCLA Institutional Review Board (IRB# 16-001 576) and the local review boards at each EpiBioS4Rx Study Group institution. Written informed consent to participate in this study was provided by the participants' legal guardian/next of kin.

To account for brain deformations and lesions due to TBI, 3D T2-FLAIR images for each patient were manually segmented using ITK-SNAP [27]. Parenchymal contusions and brain edema were segmented together as one 3D mask for this analysis. Manual segmentations were reviewed by clinicians with expertise in neuroradiology.

2.2 Preprocessing Pipelines

In order to prepare the dMRI for analysis, the acquired dMRI scans are processed in the FMRIB Software Library (FSL) [23], to estimate diffusion tensor imaging (DTI) parameters and FA images. Following the preliminary registration, the subsequent steps are carried out to extract TBSS features in four different pipelines, as outlined in Fig. 1. In Pipeline 1 and 2, no lesion information is used and are labeled as non-lesion analyses (NL). Both NL pipelines use non-linear registration, as seen in Fig. 1, to transform the FA images of individual subjects from the native space into the standard MNI space, by registering the subject image to the standard FA template (Std). In Pipeline 1, skeletonization for identifying FA values along the different WM tracts, is carried out using the mask and distance map from the Std following the method reported in [25]. In Pipeline 2, study-specific (SS) mean FA derived from the patients is used for masking individually registered FA images and skeletonization [13]. Pipelines 3 and 4 use lesion information by compensating for the T2-FLAIR lesion mask via cost function masking while registering subject space FA masks to MNI space, shown as lesion normalization (Les) in Fig. 1. Pipelines 3 and 4 extend the approach as outlined in [1]. These two Les pipelines then follow the same steps involved in the first two NL pipelines. Following skeletonization, mean FA values are calculated along a total of 63 WM tracts and WM bundles obtained from the JHU-DTI atlas [25].

2.3 Lesion Normalization

The lesion normalization technique is based on the normalization procedure given in [1], where a native space T1-weighted scan of a pathological brain is transformed into a lesion compensated image in the MNI space using a binary mask of the lesion (1 = lesion area, 0 = non-lesion area), and an MNI152 structural scan template. In this work, we have modified that technique by replacing: the T1-weighted subject scan with an FA image obtained from DTI, and the MNI152 structural scan by a standard FA template in the MNI space.

To produce the lesion mask, we first transformed the lesion mask obtained from a T2-weighted fluid attenuated inversion recovery (FLAIR) image to the MNI space and then converted it to the necessary binary mask format.

2.4 Feature Selection and Classifiers

In classification and prediction tasks, choosing a suitable classifier depends greatly on the type of input data and the desired outcome. In this work, the mean FA, taking values between 0 and 1, along the different WM pathways form a continuous-valued numerical input matrix $\mathbf{X} \in \mathbb{R}^{N \times J}$, where $N = 22$ is the number of subjects analyzed and $J = 63$ is the total number of features. The number of seizure labels (a late seizure or not) forms a binary categorical output vector $\mathbf{y} \in \mathbb{R}^{N \times 1}$. For this input-output pair, we conducted an exhaustive search among several binary classifiers of potential interest, namely: Adaboost, Random Forest, Gradient Boosting Machines, XGBoost, LDA, SVM, multi-layer perceptron (MLP) neural network, logistic regression, and Gaussian Naive Bayes.

The input feature dimension $J > N$, even without taking into consideration the reduction of training samples due to cross-validation (CV). Many of these features may add noise and hurt the overall classifier performance by increasing complexity and over-fitting [8]. It is thus recommended to select the features that will maximize the discriminative power of the classifier. Feature selection mainly takes two broad forms: wrapper methods and filter methods [8]. Wrapper methods attempt to find the optimal selection of features specifically for a given classifier, where the result of the selection is non-transferable. Conversely, filter methods dissociate the feature selection from the classifier training, by aiming to maximize some objective of similarity between the individual features and the target labels. In this work, we investigated three univariate filter methods versus no feature selection applied, on the training set of each CV fold. The univariate methods independently rank the usefulness of each feature in explaining the target label, and they are, namely: Mutual Information (MI), χ^2-test, and F-test. MI between the j-th feature and $\mathbf{y}$ is estimated using the nearest neighbor method [21]. The χ^2-test and F test are carried out in their standard formulations as implemented in the scikit-learn library [19] of Python, where the test results are looked up in their respective tables to calculate the relevant scores.

2.5 Evaluation Metrics and Experimental Details

The data is divided into a 5:1 train and test set split, following a six-fold CV. Since the data is unbalanced, a randomized stratification is included in the CV strategy, such that the percentage of samples for each class is preserved in each fold. For hyperparameter tuning, a validation set is temporarily created from only the first training set. The classifiers are then trained on the resulting reduced train set. Once the parameters for the classifiers are selected, based on the preliminary performance of the classifiers on this validation set, the CV experiments are then subsequently carried out using the entire train set.

To assess the performance of each feature selection-classifier model trained on the training set in each fold, we evaluate them over the respective test set in each of those CV folds. For each model involving a feature selection technique, the number of features selected is varied from one and up to ten features, resulting in ten repetitions of the six-fold CV. The specific features chosen, in each

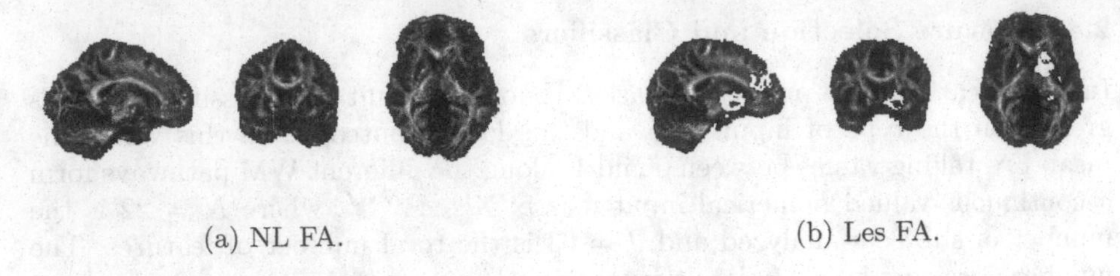

(a) NL FA. (b) Les FA.

Fig. 2. Patient 35 registered in MNI space, without and with lesion compensation.

of those ten repetitions, are essentially incorporated inside the CV procedure. Among these eleven combinations of features tested (no feature selection, and one through ten features selected), we record the number of features that lead to the best performance and the corresponding performance metrics: mean prediction accuracy, mean AUC of the receiver operating characteristic (ROC) curve, mean sensitivity, and mean specificity. The accuracy is given by

$$\text{accuracy} = \frac{\text{TP} + \text{TN}}{\text{TP} + \text{FN} + \text{TN} + \text{FP}}, \tag{1}$$

where TP, FN, TN, and FP are the total true positive, false negative, true negative, and false positive, respectively. The mean accuracy is recorded with its standard error of mean (SEM), calculated for a 95% level of confidence. The ROC analysis makes use of the sensitivity $= \frac{\text{TP}}{\text{TP+FN}}$, and the specificity $= \frac{\text{TN}}{\text{FP+TN}}$. The variation of the mean ROC curve is estimated by a binomial distribution approximation [24], for a 95% level of confidence.

The best performing model (in terms of AUC, and accuracy to tie-break if necessary) for each pipeline is identified. Since the two classes are slightly imbalanced, AUC is preferred over accuracy. In order to test for potential biomarkers, the features selected in the best performing model from each pipeline are compared between the seizure and no seizure group, with the Mann Whitney U tests performed on the entire dataset. If the null hypothesis is rejected, the feature may be recommended as a potential biomarker of late seizures.

The entire analysis is run on a workstation equipped with a 9th generation core-i7 3.6 GHz CPU, 64 GB of RAM, and an RTX 2080 Ti GPU hardware. The software used are FSL 6.0.3, and Python 3.7.4 with scikit-learn version 0.23.2 and xgboost version 1.1.1.

3 Results and Discussion

In preprocessing, the processed FA images were registered to the standard Human Connectome Project (HCP1065) DTI FA template [26]. The HCP1065 FA is slightly bigger in volume than FSL's FMRIB58 FA [23], and has a higher resolution than the ENIGMA DTI FA [25]. In preliminary experiments with dMRI scans of TBI patients, the HCP1065 FA yielded better classification performance compared to the other two templates, and was thus chosen as the Std.

Table 1. Performance comparison of the different pipelines using selected univariate feature selection methods and chosen binary classifiers.

Preprocessing	Feature selection	Classifier	Feature (s)	AUC	Accuracy ± SEM	Sensitivity	Specificity
NL-Std [Pipeline 1]	None	LDA	63	**0.718**	$\mathbf{73.6 \pm 11.8}$	**0.667**	**0.778**
	None	AdB	63	0.514	52.7 ± 20.6	0.417	0.611
	MI	LDA	3	0.635	65.3 ± 14.6	0.500	0.778
	MI	AdB	2	0.413	48.6 ± 19.7	0.500	0.583
	χ^2-test	LDA	10	0.595	59.7 ± 14.1	0.583	0.611
	χ^2-test	AdB	4	0.663	69.4 ± 13.8	0.500	0.750
	F-test	LDA	7	0.663	68.1 ± 20.5	0.667	0.667
	F-test	AdB	6	0.595	59.7 ± 17.4	0.583	0.611
NL-SS [Pipeline 2]	None	LDA	63	0.662	70.8 ± 18.0	0.500	0.833
	None	AdB	63	0.432	44.4 ± 13.1	0.333	0.528
	MI	LDA	4	**0.677**	69.4 ± 13.7	0.583	0.778
	MI	AdB	6	0.527	54.1 ± 10.7	0.500	0.556
	χ^2-test	LDA	4	0.554	54.2 ± 10.7	0.500	0.611
	χ^2-test	AdB	4	0.662	63.9 ± 14.2	0.667	0.667
	F-test	LDA	10	0.676	65.3 ± 14.6	0.583	0.778
	F-test	AdB	1	0.596	54.2 ± 15.8	0.667	0.528
Les-Std [Pipeline 3]	None	LDA	63	0.391	43.1 ± 18.2	0.250	0.528
	None	AdB	63	0.472	48.6 ± 16.1	0.417	0.528
	MI	LDA	8	0.731	73.6 ± 11.8	0.750	0.722
	MI	AdB	2	0.568	55.6 ± 6.3	0.500	0.639
	χ^2-test	LDA	6	**0.785**	$\mathbf{81.9 \pm 10.5}$	**0.750**	**0.833**
	χ^2-test	AdB	9	0.474	54.2 ± 10.7	0.250	0.694
	F-test	LDA	5	0.703	73.6 ± 16.5	0.583	0.833
	F-test	AdB	1	0.667	61.1 ± 18.7	0.649	0.639
Les-SS [Pipeline 4]	None	LDA	63	0.378	40.3 ± 18.2	0.250	0.500
	None	AdB	63	0.474	48.6 ± 11.2	0.500	0.444
	MI	LDA	3	0.622	62.8 ± 20.9	0.417	0.833
	MI	AdB	7	0.167	59.7 ± 8.1	0.500	0.833
	χ^2-test	LDA	8	0.690	69.4 ± 13.7	0.667	0.722
	χ^2-test	AdB	8	0.556	58.3 ± 12.1	0.417	0.694
	F-test	LDA	10	**0.718**	$\mathbf{73.6 \pm 11.8}$	0.667	0.778
	F-test	AdB	1	0.690	65.3 ± 18.6	0.750	0.630

The effect of the lesion normalized registration on the MNI space dMRI, when compared to a standard non-linear registration, can be seen in Fig. 2. Lesion normalization was observed to cause discontinuity of the WM tracts in the spatial location of the lesions, and the resulting images also appeared less deformed.

In our six-fold CV experiments, the Adaboost classifier consistently outperformed Random Forest, Gradient Boosting Machines and XGBoost classifiers. Among the other classifiers tested, LDA beat SVM (linear kernel), tuned MLP (two hidden layers), logistic regression, and Gaussian Naive Bayes classifiers. Since Adaboost and LDA performed the best in aggregate among all classifiers tested, we report only the performances of Adaboost and LDA. The hyperparameters tuned were the number of trees and estimators in these classifiers, while all other parameters were kept at their scikit-learn and xgboost defaults. The hyperparameters were varied among values $= \{4, 8, 16, 32, 50, 64, 100\}$, before being tuned to 100, a value that maximized the accuracy and AUC on average in the validation set.

The Adaboost and LDA classifiers were then used in combination with all the features and the three feature selection methods, for each pipeline. Results of these experiments are recorded in Table 1. We observe that both variants of the lesion normalization outperform the respective variants of the non-lesion analysis, in terms of the best seizure prediction performance by each (shown in bold). Interestingly, both standard template registration pipelines also beat their corresponding study-specific registration pipelines in performance. In particular, the Les Std (pipeline 3) with a χ^2-test based feature selection and LDA as the classifier performs the best overall. LDA emerged as the better classifier, with the best performance from each pipeline, and a faster average run-time (0.27 s) compared to Adaboost (0.97 s). The ROC plots for the best performing model from each pipeline (displayed in bold in Table 1) are shown in Fig. 3.

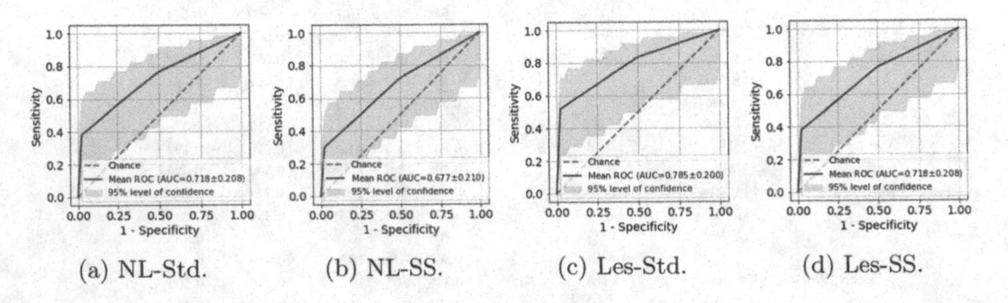

(a) NL-Std. (b) NL-SS. (c) Les-Std. (d) Les-SS.

Fig. 3. Receiver operating characteristics for the best performing models.

In order to understand whether the features that resulted in the best classification performance could also be used as potential biomarkers, we compared the distribution of the features from the best classifier in each pipeline across groups using the Mann Whitney U test. Aside from the best model in pipeline 1, each of the other models chose k features (where the value of k depended on the specific model), in each of the six folds of the CV. From the aggregate of all the six folds, we ranked the features chosen in decreasing order of their frequency of appearance. The top k most frequent features were then selected for comparison between the seizure and no seizure group. The features with uncorrected p-values less than 0.05 are reported in Table 2, along with their Bonferroni corrected p-values [7]. Table 2 shows that the mean FA in the anterior limb of the internal capsule right (ALIC-R) is significantly different between groups. White matter abnormalities of this tract have been reported earlier in patients with temporal lobe epilepsy [15]. These results provide further evidence that dMRI alterations in ALIC-R may be a potential biomarker of epileptogenesis after TBI. It is worth noting that alterations in other tracts (UNC-R, CST-L, IC-R, and PLIC-R) that we found to be significant before the Bonferroni correction have been shown to be associated with seizure propagation, duration and seizure-related WM abnormalities [5, 6, 22]. It may be that as the considered sample size increases, the effects of alterations in these tracts will become more evident.

Table 2. Comparison of features between the two seizure groups using the Mann Whitney U test, with p-values corrected using the Bonferroni correction.

Preprocessing and model	Features	Potential feature*	Test statistic	Uncorrected p-value	Corrected p-value
NL-Std [None, LDA, 63]	63	FX/ST-L	33.0	0.0475	1.000
		UNC-R	30.0	0.0308	1.000
NL-SS [MI, LDA, 4]	4	FX/ST-L	33.0	0.0475	0.1900
Les-Std [χ^2-test, LDA, 6]	6	ALIC-R	**22.0**	**0.0081**	**0.0486**
		UNC-R	30.0	0.0308	0.1848
Les-SS [F-test, LDA, 10]	10	ALIC-R	22.0	0.0081	0.0810
		CST-L	27.0	0.0192	0.1920
		IC-R	33.0	0.0475	0.4750
		PLIC-R	30.0	0.0308	0.3080
		UNC-R	31.0	0.0357	0.3570

* FX/ST-L = Fornix/stria terminalis left, UNC-R = uncinate fasciculus right, ALIC-R = anterior limb of internal capsule right, CST-L = corticospinal tract left, IC-R = internal capsule right, PLIC-R = posterior limb of internal capsule right.

4 Conclusion

In this work, we introduced cost function mapping, a lesion normalization technique, to the study of dMRI and compared various preprocessing pipelines using classification performance to judge among them. From the presented experiments, we found that the pipeline involving lesion normalization and a standard template skeletonization provides the best late seizure prediction performance. Lesion normalization shows an improvement of eight percentage points in mean accuracy and seven percentage points in mean AUC, when compared to the standard template skeletonization pipeline with non-lesion analysis. Following statistical analyses of selected features, we find evidence that the dMRI alterations of ALIC-R may serve as a biomarker of seizures post TBI.

This work has certain limitations. Since this longitudinal study is ongoing, the experiments are only carried out on a limited number of patients from the cohort, for whom the complete follow-up data has been obtained. Additionally, the mean FA values used for the analyses might not adequately characterize the spatial distribution of FA values along the tracts, where additional distribution descriptors might provide more discriminative information. Furthermore, deep convolutional models, which could provide better classification performance, have not been tested due to the limited number of samples.

For future work, we aim to expand the study by including more subjects and examine the effects of lesion normalization on dMRI registration and classification in greater detail. We would also like to explore other techniques involving the analysis of white matter tracts, and investigate other neural network classifiers that may be trained to perform well with few data samples.

References

1. FSL lesion normalization. https://neuroimaging-core-docs.readthedocs.io/en/latest/pages/fsl-anat-normalization-lesion.html
2. Study design and protocol: the epilepsy bioinformatics study for anti-epileptogenic therapy (EpiBioS4Rx) clinical biomarker. Neurobiol. Disease **123**, 110–114 (2019)

3. Akbar, M.N., La Rocca, M., Garner, R., Duncan, D., Erdoğmuş, D.: Prediction of epilepsy development in traumatic brain injury patients from diffusion weighted MRI. In: Proceedings of the 13th ACM International Conference on PErvasive Technologies Related to Assistive Environments. PETRA 2020. Association for Computing Machinery, New York (2020). https://doi.org/10.1145/3389189.3397655
4. Brett, M., Leff, A.P., Rorden, C., Ashburner, J.: Spatial normalization of brain images with focal lesions using cost function masking. NeuroImage **14**(2), 486–500 (2001)
5. Chiang, S., Levin, H.S., Wilde, E., Haneef, Z.: White matter structural connectivity changes correlate with epilepsy duration in temporal lobe epilepsy. Epilepsy Res. **120**, 37–46 (2016)
6. Destrieux, C., Velut, S., Zemmoura, I.: Symposium "brain plasticity in epilepsy", Leuven, Belgium, 13–16 May 2017. Hippocampus **59**, 149–158 (2013)
7. Dunn, O.J.: Multiple comparisons among means. J. Am. Stat. Assoc. **56**(293), 52–64 (1961)
8. Erdogmus, D., Ozertem, U., Lan, T.: Information theoretic feature selection and projection, vol. 83, pp. 1–22 (12 2007). https://doi.org/10.1007/978-3-540-75398-8_1
9. Frey, L.C.: Epidemiology of posttraumatic epilepsy: a critical review. Epilepsia **44**, 11–17 (2003)
10. Garner, R., La Rocca, M., Vespa, P., Jones, N., Monti, M.M., Toga, A.W., Duncan, D.: Imaging biomarkers of posttraumatic epileptogenesis. Epilepsia **60**(11), 2151–2162 (2019)
11. Gupta, R.K., Saksena, S., Agarwal, A., Hasan, K.M., Husain, M., Gupta, V., Narayana, P.A.: Diffusion tensor imaging in late posttraumatic epilepsy. Epilepsia **46**(9), 1465–1471 (2005)
12. Humphreys, I., Wood, R.L., Phillips, C.J., Macey, S.: The costs of traumatic brain injury: a literature review. ClinicoEcon. Outcomes Res. CEOR **5**, 281 (2013)
13. Keihaninejad, S., et al.: The importance of group-wise registration in tract based spatial statistics study of neurodegeneration: a simulation study in Alzheimer's disease. https://doi.org/10.1371/journal.pone.0045996, www.plosone.org
14. La Rocca, M., et al.: Multiplex networks to characterize seizure development in traumatic brain injury patients. Front. Neurosci. **14**, 1238 (2020)
15. Meng, L., et al.: White matter abnormalities in children and adolescents with temporal lobe epilepsy. Magn. Resonan. Imaging **28**(9), 1290–1298 (2010)
16. Morgan, V.L., et al.: Magnetic resonance imaging connectivity for the prediction of seizure outcome in temporal lobe epilepsy. Epilepsia **58**(7), 1251–1260 (2017). https://doi.org/10.1111/epi.13762
17. Nazem-Zadeh, M.R., et al.: DTI-based response-driven modeling of MTLE laterality. NeuroImage Clin. **11**, 694–706 (2016)
18. Hackenberg, K., Unterberg, A.: Der Nervenarzt **87**(2), 203–216 (2016). https://doi.org/10.1007/s00115-015-0051-3
19. Pedregosa, F., et al.: Scikit-learn: machine learning in Python. J. Mach. Learn. Res. **12**, 2825–2830 (2011)
20. Perlbarg, V., Puybasset, L., Tollard, E., Lehéricy, S., Benali, H., Galanaud, D.: Relation between brain lesion location and clinical outcome in patients with severe traumatic brain injury: a diffusion tensor imaging study using voxel-based approaches. Hum. Brain Mapp. **30**(12), 3924–3933 (2009)
21. Ross, B.C.: Mutual information between discrete and continuous data sets. PloS ONE **9**(2), e87357 (2014)

22. Schoene-Bake, J.C., et al.: Widespread affections of large fiber tracts in postoperative temporal lobe epilepsy. Neuroimage **46**(3), 569–576 (2009)
23. Smith, S.M., et al.: Tract-based spatial statistics: voxelwise analysis of multi-subject diffusion data. Neuroimage **31**(4), 1487–1505 (2006)
24. Sourati, J., Erdogmus, D., Akcakaya, M., Kazmierczak, S.C., Leen, T.K.: A novel delta check method for detecting laboratory errors northeastern university university of Pittsburgh Oregon Health & Science University Portland, OR, USA National Science Foundation, pp. 0–5 (2015)
25. Stein, J.L., et al.: Identification of common variants associated with human hippocampal and intracranial volumes. Nat. Genet. **44**(5), 552–561 (2012)
26. Yeh, F.C., et al.: Population-averaged atlas of the macroscale human structural connectome and its network topology. NeuroImage **178** (2018). https://doi.org/10.1016/j.neuroimage.2018.05.027
27. Yushkevich, P.A., et al.: User-guided 3D active contour segmentation of anatomical structures: significantly improved efficiency and reliability. NeuroImage **31**(3), 1116–1128 (2006)

Accelerating Geometry-Based Spherical Harmonics Glyphs Rendering for dMRI Using Modern OpenGL

Charles Poirier(✉), Maxime Descoteaux, and Guillaume Gilet

Department of Computer Science, Université de Sherbrooke, Sherbrooke, Canada
charles.poirier@usherbrooke.ca

Abstract. Diffusion-weighted magnetic resonance imaging is a technique aimed at measuring the displacement of water molecules inside biological tissues. From High Angular Resolution Diffusion Imaging acquisitions, it is possible to reconstruct fiber orientation distribution functions (fODF) describing the apparent quantity of white matter fibers going through a voxel for some arbitrary direction. Because these fODF are signals on the sphere, they are usually represented using a spherical harmonics (SH) basis, and visualized as radially scaled spherical glyphs. In this work, we present a novel GPU-based method for interactive real-time visualization of fODF datasets using modern OpenGL. Our algorithm relies on compute shaders to distribute the deformation of SH glyphs on the GPU, allowing real-time slicing of fODF images. We show that our method offers better performance than CPU-based methods and allows the real-time exploration of large datasets.

Keywords: Fiber ODF · Visualization · Spherical harmonics

1 Introduction

Diffusion-weighted magnetic resonance imaging (dMRI) is a technique aimed at measuring the displacement of water molecules inside biological tissues. From High Angular Resolution Diffusion Imaging [2] (HARDI), it is possible to reconstruct fiber orientation distribution functions (fODF) describing the apparent quantity [9] of white matter (WM) fibers going through a voxel for some arbitrary direction. To represent these fODF as continuous signals on the sphere, the use of a modified spherical harmonics (SH) basis has been suggested [3,16]. Because they play a major role in fiber tractography [4], it is essential to have tools to visualize and validate the reconstructed fODF [7,10]. Since it represents a signal on the sphere, it is often visualized as a radially scaled spherical glyph [11]. There are two main methods to generate this kind of glyphs.

The traditional geometry-based approach [6,12] consists in generating a sphere geometry at an arbitrary resolution and deforming it by evaluating the

Supported by NSERC Canada Graduate Scholarships - Master's Program.

S. Cetin-Karayumak et al. (Eds.): CDMRI 2021, LNCS 13006, pp. 144–155, 2021.
https://doi.org/10.1007/978-3-030-87615-9_13

signal for each sampled point on the geometry. This approach is the most common, being available in software packages such as MRtrix3 [15,17] (through the mrview visualization tool) or DIPY [5] (through the python library FURY[1]). This method is computationally expensive as it requires generating and deforming a new geometry for each voxel where a glyph will be displayed. As the deformation is often executed on the CPU, the sphere tessellation has a huge impact on the performances. In practice, sphere geometry is limited to approximately 100 vertices to ensure interactive slicing of the data. This has a negative impact on visual quality as the rendered glyphs can't capture enough details from the fODF [7]. To enhance performance and ensure interactive visualization, several optimization techniques such as generating glyphs on multiple CPU threads, levels of detail (LOD) and viewport culling have been suggested [10,11].

A second approach relies on a GPU-accelerated ray casting algorithm [1,7] to evaluate the color of each screen pixel. In this method, a ray is cast from each pixel in direction of the scene and tested against all objects. If intersections are found, the color of the closest intersection point is computed and assigned to the pixel. Since each glyph is rendered on a per pixel-basis, results are visually smoother and the method achieve high performance, although constrained by the screen resolution and the number of glyphs.

In this work, we propose to move the heavy computations required by geometry-based methods to the GPU to build a high-performance fODF visualization application using modern OpenGL. Our application allows the user to visualize three orthogonal slices loaded from a 4-dimensional SH coefficients image and supports real-time update of the slices of interest as well as basic camera movements. In [19], the authors present a GPU-based technique for visualization of Diffusion Tensor Imaging (DTI) superquadric glyphs in which they perform the sphere deformation in the vertex shader based on per-voxel parameters stored in GPU memory. They also introduce the compute shader for discriminating visible voxels. However, because the cost of evaluating a SH series is greater than the cost of evaluating superquadrics, we propose to compute the sphere deformation on a compute shader on-demand rather than on a vertex shader on a per-frame basis. We show that the use of compute shaders enables fast and efficient evaluation of SH series, unlocking new possibilities for exploring HARDI datasets in real-time. Although the technique is designed with fODF in mind, it can be used to visualize any dataset expressed in SH coefficients.

2 Methods

2.1 From SH Coefficients to Glyphs

FODF are expressed as a series of SH coefficients. To evaluate the fODF for a direction $\theta \in [0, \pi], \phi \in [0, 2\pi[$ for a given voxel, we compute the summation:

$$f(\theta, \phi) = \sum_{l=0}^{l_{max}} \sum_{m=-l}^{l} C_l^m \cdot \tilde{Y}_l^m(\theta, \phi), \qquad (1)$$

[1] https://github.com/fury-gl/fury.

where $\tilde{Y}_l^m$ is the real and symmetric modified SH function of even order l and degree m and C_l^m is a SH coefficient representing the quantity of $\tilde{Y}_l^m$ required to represent the signal. For a maximum order l_{max}, the number of SH coefficients and functions is given by

$$N = (l_{max} + 1) \cdot (l_{max} + 2)/2. \tag{2}$$

In this work, we fixed the value of l_{max} to 8, resulting in 45 SH coefficients per voxel.

Because the diffusion signal is real and symmetric by nature, we use the modified SH basis proposed by Descoteaux *et al.* [3], which has those two properties in addition to being orthonormal:

$$\tilde{Y}_l^m(\theta, \phi) = \begin{cases} \sqrt{2} \cdot Re\left[Y_l^m(\theta, \phi)\right] \; if \; m < 0 \\ Y_l^m \quad if \; m = 0 \\ \sqrt{2} \cdot Im\left[Y_l^m(\theta, \phi)\right] \; if \; m > 0 \end{cases}. \tag{3}$$

where $Re[\cdot]$, $Im[\cdot]$ respectively represent the real and imaginary part of the SH function of lth order and m^{th} degree Y_l^m, itself given by:

$$Y_l^m(\theta, \phi) = \sqrt{\frac{2l+1}{4\pi} \frac{(l-m)!}{(l+m)!}} P_l^m(\cos\theta) e^{im\phi}, \tag{4}$$

with P_l^m the associated Legendre polynomial of lth order and m^{th} degree.

To generate a glyph representing the fODF at a given voxel, we employ the geometry-based strategy mentioned above. That is, we first generate a spherical mesh of fixed resolution. Then, for each sphere direction on the mesh, we evaluate the SH series given by Eq. 1 up to order l_{max} for its corresponding (θ, ϕ) coordinates. Finally, we scale each vertex by its fODF value.

2.2 Spherical Mesh

To generate a spherical mesh, we first select a resolution R controlling the tessellation of the sphere. With $\theta \in [0, \pi], \phi \in [0, 2\pi[$, R determines the number of equally spaced points along the θ axis. Our design is such that there are twice as many points along the ϕ axis. Then, for all possible combinations of θ and ϕ, we generate a vector of unit length by converting (θ, ϕ) from spherical to Cartesian coordinates. The number of vertices for a given sphere resolution is:

$$N_{vertices} = (R - 2) \cdot 2R + 2. \tag{5}$$

Because OpenGL is designed to handle triangles, we also generate a triangulation for the generated set of vertices. The number of triangles for a set of $N_{vertices}$ points is given by:

$$N_{triangles} = 2 \cdot N_{vertices} - 4. \tag{6}$$

Although this method results in an uneven density of points throughout the mesh, we hypothesize that it is not an issue as long as there are enough points in the least dense regions of the mesh for an adequate representation of the signal. The real impact of the chosen method on the visual quality of the glyphs will be evaluated as part of future works.

2.3 Lighting

When it comes to visualizing 3-dimensional data in 2 dimensions, lighting variations play an important role as it allows to better distinguish the depth and shape of objects. As illumination models are dependant on the normal information of the surface, a normal vector needs to be computed at each vertex and interpolated across each triangle. Normals of our SH iso-surface can be computed either analytically [1,7] or numerically. In the latter case, normals are expressed on a per-vertex basis as the average of the normal of each adjacent triangle (computed as the cross-product of the triangle vectors).

Once the normals are acquired, lighting of our glyphs can be evaluated using a standard illumination model such as the popular Phong illumination [8]. We implemented the numerical approach for normals evaluation.

2.4 Compute Shaders

Compute shaders are general purpose shaders available as a OpenGL core feature since version 4.3. They enable the user to run parallel operations on GPU which are not directly related to rendering tasks. One important aspect of this type of shaders is that it can be called from anywhere in the application independently from the rendering pipeline. They can be used to execute read and write operations on Shader Storage Buffer Objects (SSBO), memory blocks allocated on the GPU which are accessible from any pipeline stage. In this work, we propose to use the compute shader to evaluate our SH series for each voxel of interest and to store the resulting vertices and normals in SSBO which will later be accessed by the rendering pipeline.

2.5 Arrays Indexing

The input data is a 4-dimensional volume of SH coefficients. However, because the OpenGL shader language (GLSL) does not handle multidimensional arrays of variable length efficiently, we express our 4-dimensional volume as a 1-dimensional array. Once flattened, the coefficient C_l^m at voxel position (v_x, v_y, v_z) is bound to index j by the following equation:

$$j = (l(l+1)/2 + m)D_x D_y D_z + v_z D_x D_y + v_y D_x + v_x, \tag{7}$$

where D_x, D_y, D_z represent the dimensions of the voxel grid along the x, y, z axes respectively.

The glyphs to display are also stored in a 1-dimensional array. Their order is such that the first glyphs belong to the Z slice, followed by the X slice and finally the Y slice. This ordering is illustrated in Fig. 1. To recover the grid position of a glyph from its flattened index, the grid dimensions are used to determine on which plane it belongs. It is thus trivial to determine its corresponding 3-dimensional position using the dimensions of the current plane and its slice index.

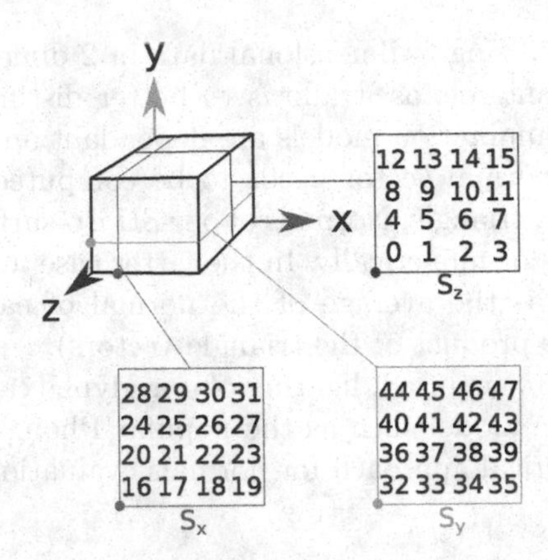

Fig. 1. Illustration of glyphs indexing inside three orthogonal slices S_x, S_y, S_z for a $4 \times 4 \times 4$ volume. The blue, red and green dots identify the bottom left corner of the three slices. The corner of the Z slices is occluded by the volume. (Color figure online)

2.6 Implementation

We designed an interactive visualization application for Linux using C++17 and OpenGL 4.6. The user specifies an SH image to visualize and the program displays SH glyphs for three orthogonal slices of the volume—one plane along each axis. The resolution of the sphere used for rendering can also be specified as an argument. The user can directly interact with the scene with the mouse for rotating, scaling and translating the model in real-time. Through a graphical user interface (GUI), the user can also control the indices of the slices to render, a multiplicative scaling factor for all glyphs and a threshold to apply on the coefficients C_0^0. The pipeline is divided between the CPU, where the data is prepared and user inputs are handled, and the GPU, where SH glyphs are instantiated for the slices of interest based on SH coefficients and rendered.

CPU Preprocessing. On the CPU, we first generate a template unit sphere of a given resolution R. For each of the sphere directions, we precompute the associated SH functions using Eq. 3.

Because we have on one hand data concerning voxels—the SH coefficients image—and on the other hand data concerning glyphs to display—all the vertices and normals of the glyphs contained in our three orthogonal slices, as well as the draw parameters for handling each one of them—we distribute their initialization on two CPU threads. This way, one thread generates the flattened array containing all SH coefficients, while the other thread generates two zero-filled arrays v_{all} and n_{all} for respectively the vertices and the normals. The number of glyphs is given by

$$N_{glyphs} = D_x D_y + D_x D_z + D_y D_z, \quad (8)$$

with D_x, D_y, D_z the grid dimensions. Therefore, v_{all} and n_{all} each contain N_{glyphs} times $N_{vertices}$. Once both threads have finished executing, all generated data is transferred on the GPU in SSBO.

During the rendering loop, all glyphs are rendered using a single OpenGL call (`glMultiDrawElementsIndirect`) to maximize the performance. The compute shader pass generating slices data is launched on demand, *i.e.* if an update of the glyphs is required.

GPU Processing. The GPU handles the scaling of the glyphs and the rendering of the scene. The scaling of glyphs is done on the compute shader, where each thread is responsible for deforming one sphere. A unique identifier I_{glyphs} in the range $\{0..N_{glyphs}\}$ determines which sphere a thread should process. From this identifier, a glyph's corresponding grid position and associated SH coefficients can be retrieved. Using the template sphere, $f(\theta, \psi)$ is evaluated with Eq. 1 using the precomputed SH functions. The scaled vertices are stored as $N_{vertices}$ consecutive vectors in v_{all}. The normals evaluation is only done when the resulting glyph has a L1-norm greater than 0 to avoid unnecessary processing. When it is the case, the faces normals are computed, smoothed and stored in a per-vertex basis.

The compute shader pass yields the two arrays v_{all} and n_{all} filled with the glyphs vertices positions and normals. Whenever the user makes changes impacting the aspect of the glyphs, such as changing slice or modifying the C_0^0 threshold, a compute shader pass is launched. When changing a slice, only the glyphs belonging to the moving slice are evaluated to avoid useless operations.

Finally, the rendering is straightforward; for each glyph instance, its position and normal in world coordinates is computed and colored accordingly. Each vertex to draw is identified by a vertex identifier corresponding to its index in v_{all} and n_{all}. The vertex color is the absolute value of its normalized direction. To express it in world coordinates, we simply translate it by its corresponding grid position using I_{glyphs}. It is also at this stage that the scaling factor is applied to the vertices, so that changing the scaling does not require an execution of the compute shader.

The source code is freely available on Github[2]. The application can be easily installed using CMake and does not depend on any external libraries, as third party libraries are already included in the project.

[2] https://github.com/CHrlS98/RTfODFSlicer.

2.7 Experiments

To evaluate our proposed method, we reconstructed fODF for a subject from the Human Connectome Project [18] (HCP) using Tractoflow [13]. The fODF reconstruction was done using constrained spherical deconvolution (CSD) [14] as implemented in DIPY with a maximum SH order of 8. The resulting image is a 4-dimensional volume of dimensions $105 \times 138 \times 111 \times 45$. Displaying three slices for this volume results in 41463 glyphs. To give an idea of the size of the dataset, a screenshot for three orthogonal slices is shown in Fig. 2. The tests were run on a notebook computer with a Intel Core i7 2.60 GHz processor, 16 GB RAM and nvidia GeForce GTX 1660Ti 6 GB GPU. We computed the average time required to update a slice for various sphere resolutions by automatically incrementing the slice index along the X axis (of dimensions 138×111) at each frame, triggering an execution of the compute shader each time. The mean update time required in FURY to display a new slice using the ODF slicer actor for default DIPY spheres of 100, 724 and 1442 vertices was evaluated using the same method. We also measured the time required for updating a slice of interest in MRtrix3's mrview for sphere LOD of 1, 2 and 3, corresponding to 66, 258 and 1074 vertices respectively. To render the same number of glyphs in all experiments, we tested mrview in *Ortho view*. Because we do not have an implementation of the ray-casting method proposed in [1] at our disposal and because the results from the original paper are reported for hardware with outdated specifications, a direct comparison with our method is not possible. We also reported the average frame rate for rendering with our method, that is without modifying glyphs between frames for different sphere resolutions. Finally, we evaluated qualitatively the visual quality of our glyphs by comparing a region of interest for various sphere resolutions.

3 Results and Discussion

Figure 3 illustrates the relationship between the average time required to update glyphs and the sphere resolution for a slice of 138×111 (15318 glyphs) along the X axis for our method, a FURY-based implementation and MRtrix3. In all cases, the relationship is linear. This is expected because an increased number of vertices results in an increased number of spherical functions to evaluate. Using a sphere of 162 vertices, we report an average update time of approximately 0.027 s using our implementation. Using our FURY-based application with a sphere of 100 vertices, the average update time is of 0.45 s. For a sphere LOD of 1 in mrview, resulting in 66 vertices per sphere, the average update time is of 0.294 s. Comparing the curves' slope, our method scales better than FURY and similarly to MRtrix3 relative to the number of vertices. From these results, we can say that our method is at least 17 times faster than a FURY-based method for any sphere of more than 100 vertices and at least 10 times faster than MRtrix3's implementation. However, it is important to note that these comparisons, while informative, are not especially fair. For example, mrview's update time depends

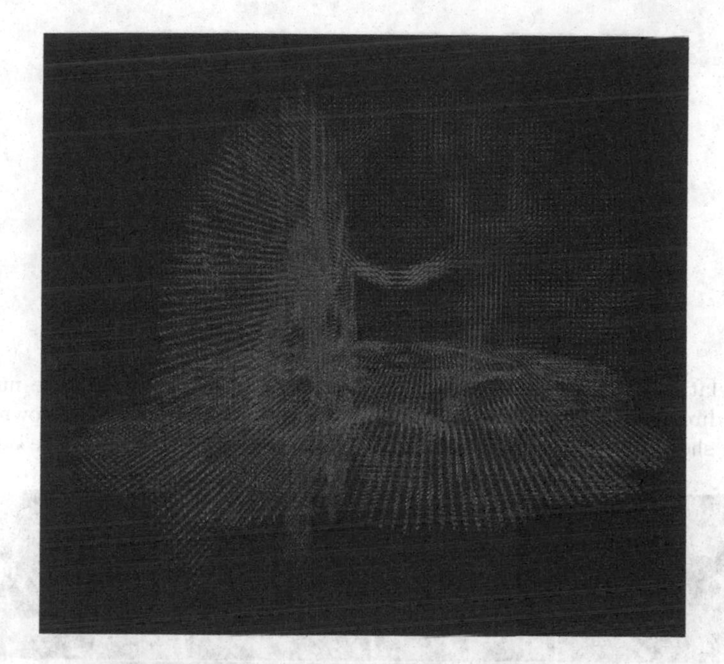

Fig. 2. Overview of the dataset used for testing. The screenshot is taken from our visualization application.

on the number of visible objects whereas our method does not. Therefore, when zooming in a region of interest, mrview can be faster than our application. Also, when displaying only one slice in mrview, we report a lesser slope than using our application (single slice visualization is not yet supported), meaning MRtrix3 scales better than our application in this situation.

The Fig. 4 shows the effect of an increasing sphere resolution on the visual quality of a glyph taken from a region of interest for sphere resolutions of 5 to 35. For small sphere resolutions, we see that resulting glyphs display very sharp edges and lack detail to allow a sufficient representation of the underlying signal. The lighting also suffers from an insufficient resolution because the normal varies too much between neighbour faces. For resolution 15 and above, the glyphs better capture the shape of the signal. We also report little improvement in visual quality between spheres of resolution 25, 30 or 35.

The Table 1 shows the impact of the number of vertices and triangles per sphere on the frame rate of the application. These values are measured when navigating through the scene without modifying the aspect of the glyphs between two frames. Using our test dataset, there are $41463 \cdot N_{triangles}$ drawn at every frame. For example, with a sphere of resolution 25, our application renders 95364900 triangles, 48 times per seconds. Thanks to our efficient OpenGL draw calls, we report frame rates high enough for a smooth navigation for all tested values, allowing the interactive visualization of complex datasets.

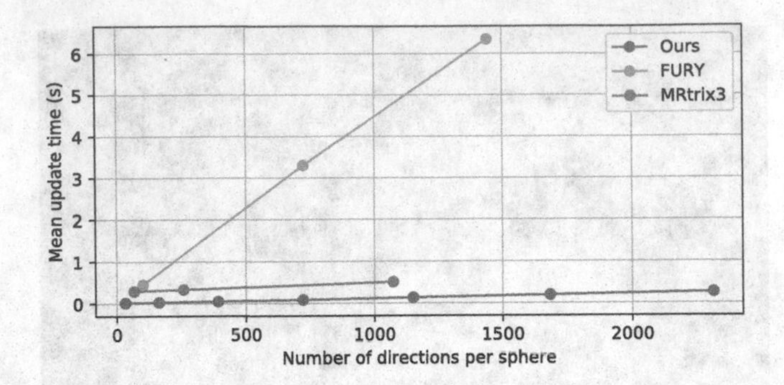

Fig. 3. Relationship between the average update time per slice and the number of evaluated directions per sphere. Times measured using our method are shown in blue, FURY are shown in orange and MRtrix3 are shown in green. (Color figure online)

Fig. 4. Effect of the sphere resolution on visual quality of the glyphs for a region of interest. From top-left to bottom-right are shown sphere resolutions going from 5 to 35 by steps of 5.

Table 1. Mean frame rate for various sphere resolutions

Resolution	Number of vertices	Number of triangles	Frame rate (fps)
5	32	60	1263
10	162	320	371
15	392	780	133
20	722	1440	74
25	1152	2300	48
30	1682	3360	33
35	2312	4620	24

By taking into account all the previous measurements, we determine that a sphere resolution of 25 is a satisfying trade-off between performances and visual quality. Figure 5 shows a snapshot of fODF in the centrum semiovale for a coronal slice in which we can appreciate nicely defined two- and three-fibers crossings.

Fig. 5. Snapshot of fODF in centrum semiovale for a coronal (Y) slice where we can clearly distinguish two- and three-fibers crossings.

As a limitation, we note that, because the draw call used in this work is not available for lower OpenGL versions, our software is limited to GPU hardware supporting OpenGL 4.6. A less efficient draw call would have to be used in order to make the application available to a wider range of GPU and, ultimately, a CPU implementation of the sphere deformation would be needed in order to support very old hardware or without GPU capabilities. However, the performance of our proposed method is dependent on the GPU processing power and will scale accordingly to the GPU capabilities.

4 Conclusion

In the present work, we proposed a novel method for generating and visualizing glyphs from a signal expressed as a series of SH coefficients using state-of-the-art shader technology. We took advantage of modern OpenGL compute shaders to distribute the evaluation of the SH coefficients series on the GPU for an important quantity of voxels. We report both excellent performances and excellent visual quality. Indeed, our method is at least 17 times faster at updating a 138×111 slice with glyphs of 100 vertices and above compared to a FURY-based application and at least 10 times faster than MRtrix3's implementation for LOD of 1 to 3 in *Ortho view*. To obtain better performances, we could integrate LOD [10,19] based on the distance between the user and the glyphs in order to display closer glyphs with a higher resolution than glyphs that are far away. These LOD could even be used for displaying information of different nature based on the distance at which the data is looked at, with high-level scalar maps being displayed when the user is far and high-resolution SH glyphs when closer. We could also add the possibility to magnify any glyph of interest, which could be visualized and rotated in a smaller viewport in the application window. We would also like to add the option to visualize 2-dimensional scalar maps behind the SH glyphs for the slices of interest like in MRtrix3. Additionally, we have plans to extend our application for supporting additional bases for representing fODF. We hope these future improvements could give the user a better insight on the nature of the data. Finally, the framework implemented in this work could also serve as a basis for studying the effect of novel fODF field filtering algorithms in real-time, allowing interactive configuration of filter parameters and instantaneous feedback.

References

1. van Almsick, M., et al.: GPU-based ray-casting of spherical functions applied to high angular resolution diffusion imaging. IEEE Trans. Visual. Comput. Graph. **17**(5), 612–625 (2011). Conference Name: IEEE Transactions on Visualization and Computer Graphics. https://doi.org/10.1109/TVCG.2010.61
2. Descoteaux, M.: High angular resolution diffusion MRI: from local estimation to segmentation and tractography. Ph.D. thesis, Université Nice-Sophia Antipolis, February 2008
3. Descoteaux, M., Angelino, E., Fitzgibbons, S., Deriche, R.: Regularized, fast, and robust analytical Q-ball imaging. Magn. Resonan. Med. **58**(3), 497–510 (2007). https://doi.org/10.1002/mrm.21277
4. Descoteaux, M., Deriche, R., Knösche, T.R., Anwander, A.: Deterministic and probabilistic tractography based on complex fiber orientation distributions. IEEE Trans. Med. Imaging **28**(2), 269–286 (2009). Publisher: Institute of Electrical and Electronics Engineers. https://doi.org/10.1109/TMI.2008.2004424
5. Garyfallidis, E., et al.: Dipy, a library for the analysis of diffusion MRI data. Front. Neuroinform. **8** (2014). Publisher: Frontiers. https://doi.org/10.3389/fninf.2014.00008
6. Hlawitschka, M., Scheuermann, G.: HOT-lines: tracking lines in higher order tensor fields. In: VIS 2005. IEEE Visualization, 2005, pp. 27–34, October 2005. https://doi.org/10.1109/VISUAL.2005.1532773
7. Peeters, T., Prckovska, V., van Almsick, M., Vilanova, A., ter Haar Romeny, B.: Fast and sleek glyph rendering for interactive HARDI data exploration. In: 2009 IEEE Pacific Visualization Symposium, pp. 153–160, April 2009. ISSN 2165-8773. https://doi.org/10.1109/PACIFICVIS.2009.4906851
8. Phong, B.T.: Illumination for computer generated pictures. Commun. ACM **18**(6), 7 (1975)
9. Raffelt, D., et al.: Apparent fibre density: a novel measure for the analysis of diffusion-weighted magnetic resonance images. NeuroImage **59**(4), 3976–3994 (2012). https://doi.org/10.1016/j.neuroimage.2011.10.045
10. Schultz, T., Kindlmann, G.: A Maximum enhancing higher-order tensor glyph. Comput. Graph. Forum **29**(3), 1143–1152 (2010). https://doi.org/10.1111/j.1467-8659.2009.01675.x
11. Schultz, T., Vilanova, A.: Diffusion MRI visualization. NMR Biomed. **32**(4), e3902 (2019). https://doi.org/10.1002/nbm.3902
12. Shattuck, D.W., et al.: Visualization tools for high angular resolution diffusion imaging. In: Metaxas, D., Axel, L., Fichtinger, G., Székely, G. (eds.) MICCAI 2008. LNCS, vol. 5242, pp. 298–305. Springer, Heidelberg (2008). https://doi.org/10.1007/978-3-540-85990-1_36
13. Theaud, G., et al.: TractoFlow: a robust, efficient and reproducible diffusion MRI pipeline leveraging Nextflow & Singularity. NeuroImage **218**, 116889 (Sep 2020). https://doi.org/10.1016/j.neuroimage.2020.116889

14. Tournier, J.D., Calamante, F., Connelly, A.: Robust determination of the fibre orientation distribution in diffusion MRI: non-negativity constrained super-resolved spherical deconvolution. NeuroImage **35**(4), 1459–1472 (2007). https://doi.org/10.1016/j.neuroimage.2007.02.016
15. Tournier, J.D., Calamante, F., Connelly, A.: MRtrix: diffusion tractography in crossing fiber regions. Int. J. Imaging Syst. Technol. **22**(1), 53–66 (2012). https://doi.org/10.1002/ima.22005
16. Tournier, J.D., Calamante, F., Gadian, D.G., Connelly, A.: Direct estimation of the fiber orientation density function from diffusion-weighted MRI data using spherical deconvolution. NeuroImage **23**(3), 1176–1185 (2004). https://doi.org/10.1016/j.neuroimage.2004.07.037
17. Tournier, J.D., et al.: MRtrix3: a fast, flexible and open software framework for medical image processing and visualisation. NeuroImage **202**, 116137 (2019). https://doi.org/10.1016/j.neuroimage.2019.116137
18. Van Essen, D.C., et al.: The WU-Minn human connectome project: an overview. NeuroImage **80**, 62–79 (2013). https://doi.org/10.1016/j.neuroimage.2013.05.041
19. Voltoline, R., Wu, S.T.: Multimodal visualization of complementary color-coded fa map and tensor glyphs for interactive tractography ROI seeding. Comput. Graph. **96**, 24–35 (2021). https://doi.org/10.1016/j.cag.2021.03.001

DiSCo Challenge - Invited Contribution

The Microstructural Features of the Diffusion-Simulated Connectivity (DiSCo) Dataset

Jonathan Rafael-Patino[1,2](✉), Gabriel Girard[1,2,3], Raphaël Truffet[4], Marco Pizzolato[1,5], Jean-Philippe Thiran[1,2,3], and Emmanuel Caruyer[4]

[1] Signal Processing Lab (LTS5), École Polytechnique Fédérale de Lausanne (EPFL), Lausanne, Switzerland
jonathan.patinolopez@epfl.ch
[2] Radiology Department, Centre Hospitalier Universitaire Vaudois (CHUV), University of Lausanne (UNIL), Lausanne, Switzerland
[3] Center for BioMedical Imaging (CIBM), Lausanne, Switzerland
[4] Univ Rennes, Inria, CNRS, Inserm, IRISA UMR 6074, Empenn ERL U-1228, 35000 Rennes, France
[5] Department of Applied Mathematics and Computer Science, Technical University of Denmark, Kongens Lyngby, Denmark

Abstract. We present a detailed description of the structural characteristics of the MICCAI 2021 Diffusion Simulated Connectivity (DiSCo) Challenge synthetic dataset. The DiSCo dataset are one of a kind numerical phantoms for the simulation of the diffusion-weighted images (DWIs) via Monte-Carlo diffusion simulations. The microscopic and macroscopic complexity of the synthetic substrates allows the evaluation of processing pipelines for the estimation of the *quantitative* structural connectivity. The diffusion-weighted signal in each voxel of the DWIs is obtained from Monte-Carlo simulations of particle dynamics within a substrate of an unprecedented size of 1 mm^3, allowing for an image matrix size up to $40 \times 40 \times 40$ voxels (isotropic voxel sizes of 25 μm). In this paper, we provide a characterization of the microstructural properties of the DiSCo dataset, which is composed of three numerical phantoms with comparable microstructure. We report the ground-truth tissue volume fractions ("intra-axonal", "extra-axonal", "myelin"), the fibre density, the bundle density and the fibre orientation distributions (FODs). We believe that this characterization will be beneficial for validating quantitative structural connectivity processing pipelines, and that could eventually find use in microstructural modelling based on machine learning approaches.

Keywords: Monte-Carlo simulations · DW-MRI · Phantoms · Tractography · Microstructure

J. Rafael-Patino and G. Girard—These two authors contributed equally.

S. Cetin-Karayumak et al. (Eds.): CDMRI 2021, LNCS 13006, pp. 159–170, 2021.
https://doi.org/10.1007/978-3-030-87615-9_14

1 Introduction

For the last two decades, diffusion-weighted magnetic resonance imaging (DW-MRI) has been an active area of research, with numerous contributions to the development of structural connectivity analyses. However, it is difficult to quantify the effect of a particular element of the DW-MRI data processing pipeline, like noise reduction methods [9,29], local reconstruction methods of the angular diffusion information [26,30], or tractography algorithms [12,27], on the structural connectivity results. Furthermore, in order to obtain a quantitative comparison of these methods, the use of tracers on animal models [18], or post-mortem dissection, or cortical electro-stimulation is required [19]. These techniques are time-consuming and moderately to highly invasive, and they do not provide a systematic ground truth mapping of the axonal fibre pathways.

To overcome such challenges, some physical phantoms have been developed [16,19], providing a convenient way to evaluate DW-MRI image processing methods in a more quantitative manner. However, these phantoms' fibre geometries and microstructural features are typically much simpler than those found in the brain. Moreover, the precise structural measurements of the manufactured phantom may be not fully known, defeating the purpose of using such phantoms. Numerical phantoms are of particular interest in this context and have become the standard *de facto* for evaluating novel DW-MRI signal processing methods [1,4,6,7,13,14,21]. The realism of phantoms is a fundamental aspect. Generally, it is possible to think of two levels of realism connected to numerical phantoms for DWIs. One, *macroscopic*, has to do with the fidelity to the known key features of the tissue organization, such as the complex and convoluted trajectories of white matter fibres and their configuration. The other, *microscopic*, is the fidelity of the numerical phantoms to the potential properties of the tissue microstructure such as its composition—axons, myelin, etc.—, geometrical features—axonal radii—, and physicochemical characteristics that are relevant for characterizing the tissue magnetization—such as the transverse relaxation time.

Freely available software have been developed and released [7,8,17] to create numerical phantoms for validating structural connectivity pipelines. For instance, Phantomas [7] and Fiberfox [17] allow the creation of complex DW-MRI signal from user-defined fibre configurations and diffusion parameters. Additionally, the Numerical Fibre Generator (NFG) [8] framework generates numerical structures randomly, resulting in an intricate set of fibre bundles from which DW-MRI images are generated. While these methods are capable of generating DWIs from substrates containing a large number of bundles of axonal fibers, they fall short on the microscopic realism that is necessary for evaluating a more *quantitative* structural connectivity.

The fidelity to the microstructural properties of the white matter tissue can be achieved with Monte-Carlo Diffusion Simulation (MCDS). In contrast with the approaches mentioned before, MCDS does not require an explicit model of the diffusion signal. Instead, MCDS requires a precise physical representation of the tissue geometry in the form of a 3D mesh substrate used to generate the dynamics of virtual water particles diffusing within and interacting with the substrate's barriers. MCDS is known for being computationally expensive and time-consuming. Moreover, it requires careful setup of the simulation parameters, design of the 3D mesh substrates, and handling of the particle interactions. In recent years a notable effort has been made to introduce state-of-the-art methods to obtain faster and more robust simulations [21,31], as well as state-of-the-art frameworks to create complex mesh substrates [6,13]. However, the computational expensiveness of these methods has still limited its use to single-voxel simulations, away from the demands of connectome validation studies.

In the context of macro- and microscopically realistic simulations, and in an effort to provide means for jointly evaluating local reconstruction, tractography, and connectivity methods, we developed the DiSCo dataset, a Monte-Carlo based dataset of unprecedented complexity and volumetric size. The numerical DiSCo phantoms are large enough ($1\,\mathrm{mm}^3$) to test tractography and connectivity methods, while also having rich microstructural properties suitable for testing tissue biophysical modeling and orientation estimation methods. In this work, we present a detailed analysis of the MICCAI 2021 DiSCo challenge numerical phantoms, reporting ground-truth microstructural maps at various resolutions, such as the voxel-wise fibre orientation distributions, the compartmental volume fractions and fibre density, and the mean axon diameter distribution.

2 Methods

The three phantoms shown in Fig. 1 (coined as DiSCo1, DiSCo2 and DiSCo3), were constructed following the procedure described on [20] using 16 randomly generated regions of interest (ROIs). The ROIs are then used to generate a connectome with desired properties, like sparsity, weight randomness and non-self connections. The main differences between the phantoms arise from the randomly generated ROIs and from the set of randomly generated non-zeros weights defining the weighted connection between them. However, due to the strands optimization procedure based on the NFG to pack and interdigitate the generated strands connecting the ROIs, structural differences are introduced in terms of the resulting number and orientations of the axons' bundles per voxel, effective diameter distribution, and compartmental volume fractions. Some of these differences are known *a priori* from the ground truth information used for the design of the phantoms, however, due to the complexity of the resulting substrate, some other features need to be estimated after the phantom has been produced.

The phantoms contain three water tissue compartments, intra-axonal, extra-axonal and myelin. The signal was simulated separately for each compartment using the MC/DC simulator [21] using the settings described in [20]. The myelin compartment was simplified as a non-diffusing compartment with water fraction proportional to the myelin volume. All the maps we report were computed using the strands' information generated from the final meshing procedure [20] in which an inner and outer layer was added as follows. The **strands** are defined

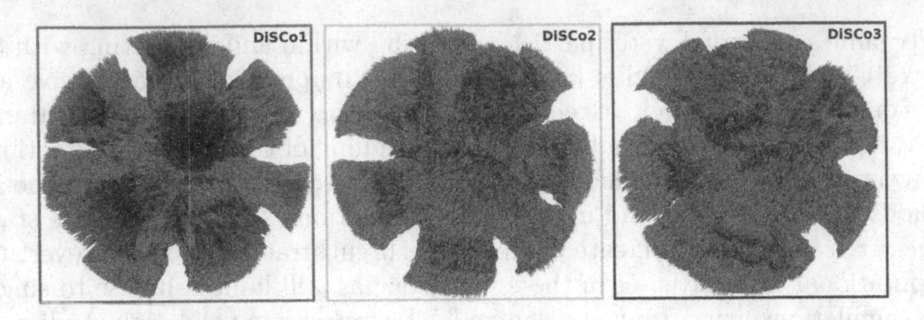

Fig. 1. Meshes of the three phantoms (DiSCo1, DiSCo2, DiSCo3) obtained following the strand optimization procedure [20]. The strands have their endpoints on the surface a sphere and trajectories propagating inside the sphere. Each strand interconnects two of the 16 ROIs.

by its center-line and the cross-sectional area, which the are used to construct the **outer and inner mesh** given the strands trajectories. The outer mesh is defined using the strands cross-sectional diameter, from which an inner mesh is generated using a down-scaled diameter by a 0.7 factor (considered as the g-ratio). The **bundles** are then defined as the set of strands that starts and ends in the two specific ROIs.

2.1 Volume Fraction Estimation

The compartmental volume fractions were computed via the Monte-Carlo sampling procedure of the diffusion simulations. In order to do so, we tracked the position of each individual *i-th* particle at time 0, $p_{i,0}$, and evaluated to which compartmental domain $\Omega \in \mathbb{R}^3$ that position belongs. In particular, we defined the intra-axonal compartment, $\Omega_i; \Omega \subset \Omega_i$, as any enclosed domain with no other substrate elements inside; the outer axonal-space (Ω_o)—related to a specific subspace Ω_i—was defined then as any enclosed domain fully containing the subspace of the intra-axonal subspace Ω_i. With this, the compartmental myelin volume fraction can be defined as the space in between those two, $\Omega_m = \Omega_o - \Omega_i$. Finally, we defined the extra-axonal compartment as anything else outside the outer compartment ($\Omega_e = \Omega - \Omega_o$). The final volume fractions maps were computed by uniformly sampling the substrate space Ω with a particle density of one particle per μm^2. The volume fraction maps were computed by subdividing the averages into the voxel regions using the maximum resolution grid of $25 \times 25 \times 25\,\mu m^2$.

2.2 Fibers Information Maps

The ground truth fibre orientation distribution functions (FODs) were computed using the strand trajectories and the cross-sectional areas. The FOD in a particular voxel is estimated from a collection of directions, representing the variability of the fibre directions within that voxel. This accounts for the different fiber bundles potentially passing through but not wholly contained in the voxel, the diameters of the fibers, and for the angular dispersion of bending strands. In order to translate this discrete representation of the FOD into a continuous representation, we used kernel density estimation (KDE), using a symmetric Von-Mises Fisher kernel, defined as

$$v_{\mu}(\omega) = c_{\kappa} \exp(\kappa|\mu^{\mathrm{T}}\omega|), \qquad \text{where} \qquad c_{\kappa} = \frac{\kappa}{4\pi(\exp(\kappa)-1)}, \tag{1}$$

where κ is the concentration parameter, and μ is the axis. This function has already been used in the context of diffusion MRI modelling [7,15,32]. The FOD in each voxel is obtained by summing the kernel aligned with μ for all fibre segments intersecting the voxel. Moreover, each kernel is weighted by the length of the segment and by its cross-sectional area to account for various fiber volumes. The diameter of the circumference defining the cross-sectional area is used to compute the effective axon diameter distribution map per voxel. The number of strands and bundles per voxel were also computed for three nominal resolutions (25 µm, 50 µm and 100 µm) using the weighted approach of the FOD explained before being separated by strand or by ROI bundle.

2.3 Peaks Extraction

Peaks were extracted from the FODs for each voxel size (25 µm, 50 µm and 100 µm) using Dipy [12]. Peaks were kept only if the FOD value in the peak orientation was equal or more than 20% of the FOD maximum (relative_peak_threshold = 0.2). The minimum separation angle between peaks was set to 25° (if multiple peaks are identify within a 25° angle, the peak with the highest FOD value is kept).

2.4 Tensor-Based Metrics

The diffusion tensors [5] and the corresponding fractional anisotropy (FA) and mean diffusivity (MD) maps were computed using the re-weighted least squares method implemented in MRtrix3 [28]. They were estimated using the full noiseless DW-MRI signal [20].

3 Results

3.1 Compartmental Volume Fraction

Figure 2 shows a cut section of the estimated ground-truth map of DiSCo1 for the three resolutions. The extra-axonal space is shown in the first row, which also contains the free water outside the main phantom corpus. The maps show the close relationship between the intra-axonal and the myelin fraction (bottom row). For the highest resolution, the combined volume fraction of these compartments is about 52% in the highly dense areas near the main center area of the phantom. Such value may result in a less hindered extra-axonal compartment compared

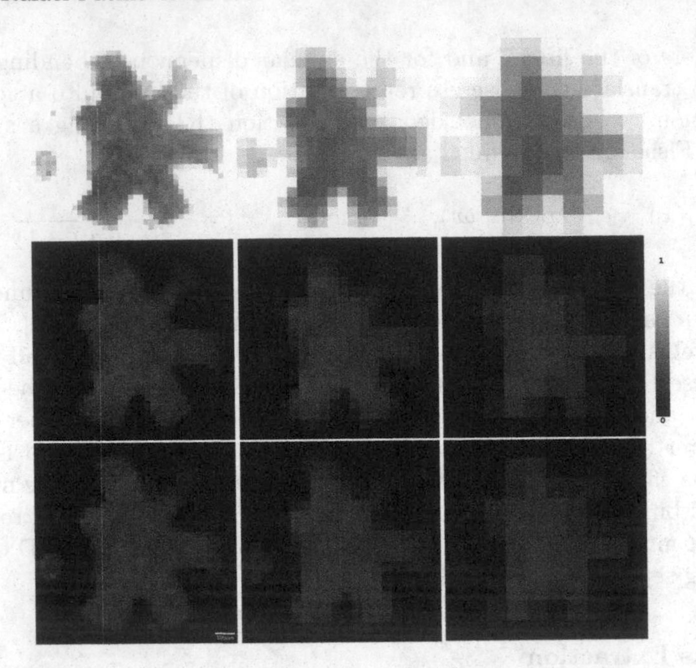

Fig. 2. Ground truth volume fraction map of DiSCo1 of the extra-axonal compartment (top), intra strand compartment (middle) and myelin layer compartment (bottom). The voxel size of the image voxel size was set to 25 μm (left), 50 μm (center) and 100 μm (right) isotropic.

to that expected in real tissue. In the lower resolution, this value can be even smaller since the partial volume is present in most of the voxels.

Figure 3 shows the histogram of the volume fractions on the three phantoms and for the three resolutions. The effect of the partial volume in the compartments' volume fractions is noticeable especially starting from 50 μm isotropic resolution.

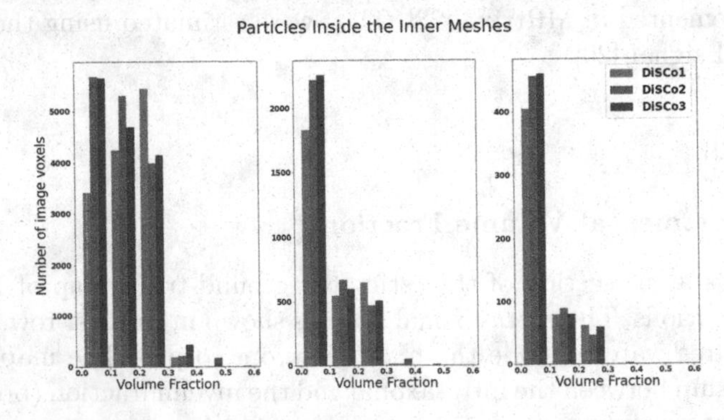

Fig. 3. Histograms of the fraction of inner strand fraction for the phantoms DiSCo1 (red), DiSCo2 (green) and DiSCo3 (blue). The voxel size of the image voxel size was set to 25 μm (left), 50 μm (center) and 100 μm (right) isotropic. (Color figure online)

3.2 FODs and Number of Streamlines as a Function of Resolution

The top-row images of Fig. 5 show the voxel-wise count of the number of strands for a section of the DiSCo1 phantom. As expected, the number of fibers is higher as the resolution decreases. At the highest resolution the maximum number of strands in a single voxel is 82 and the maximum number of bundles is 5. Conversely, at the lowest resolution the maximum number of strands in a single voxel is 1136, and 18 is the maximum number of different bundles. At the highest resolution the voxel-wise mean diameter ranges from 1.3 µm to 4.5 µm, centered at 2.25 µm, which is comparable to the range of values at the other two resolutions.

The ground truth orientations and number of peaks of a cross section of DiSCo1 is shown in Fig. 4 for the various resolutions. Bundles close to the ROIs are notably more homogeneous than those in crossing areas, which can also be noted in the FA maps in Fig. 6. The number of peaks in a single voxel is shown in the second row; notably, some highly dense voxels contained a total of 8 peaks in the FOD beyond the set threshold (see Methods section).

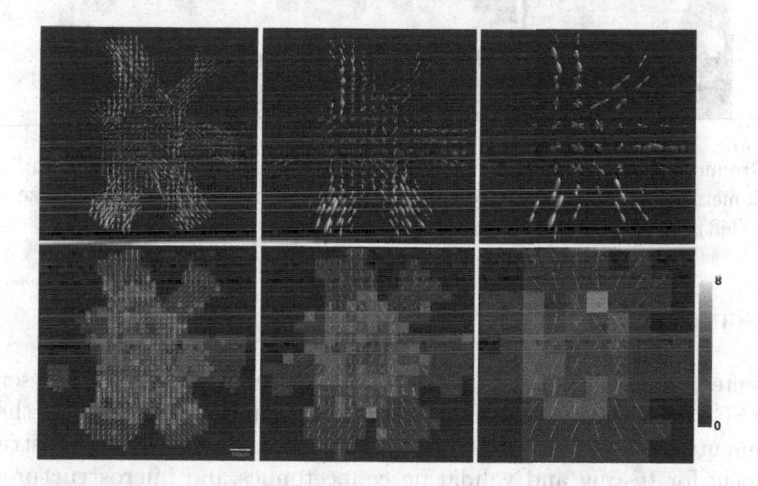

Fig. 4. Ground truth fibre orientation distribution functions (top) and corresponding peaks (bottom). The peaks are overlaid onto the peak count map. The voxel size of the image voxel size was set to 25 µm (left), 50 µm (center) and 100 µm (right) isotropic.

3.3 MD and FA Maps

The Diffusion Tensors (DT) derived maps are shown in Fig. 6. In the top row, the resulting DT maps are shown. The effect of partial volume in the lowest resolution is particularly evident in the DT maps, where single bundles near the ROIs may look fully anisotropic and thus have higher FA (as shown in the second row). The mean diffusivity is shown in the bottom row. From these maps, it is possible to observe that in the correspondence of the crossing area, the mean diffusivity is still remarkably low and homogeneous despite having a high extra-axonal volume fractions and tortuous structure.

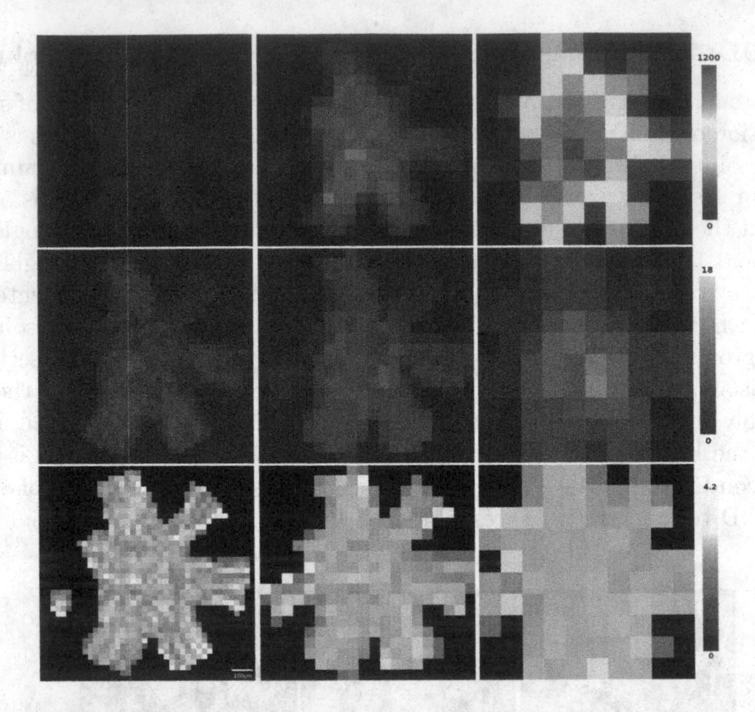

Fig. 5. Ground truth strand count map (top), bundle count map (middle) and average strand diameter map in μm (bottom). The voxel size of the image voxel size was set to 25 µm (left), 50 µm (center) and 100 µm (right) isotropic.

4 Discussion and Conclusion

We presented quantitative maps of the microstructural properties representative of the DiSCo phantoms. These maps show the complexity achieved in the three main computed resolutions and provide a novel and multiplex microstructural environment for testing and validating connectomics and microstructural techniques. For instance, besides the context of connectomics analysis, which was the focus of the DiSCo 2021 Challenge, these phantoms can be used for validating dispersion based techniques [32], multi-tensor approaches [23,25], axons diameter mapping [2], acquisition strategies for tractography [24], and tractogram filtering methods [10]. Secondly, these phantoms can be used as well to test or train DWI-based super-resolution approaches [3] given the availability of the three different resolutions presented here. However, from our experiments, we noted that the fidelity to the microstructure at the lowest resolution might be too poor and suffer from excessive partial volume effects. Another important factor to consider is the availability of two additional phantoms which can be used as test and validation datasets as classically needed for machine learning approaches. We verified in our experiments that the framework can create distinct connectomes while preserving the microstructural coherence, like the volume preservation, and achieving diffusion characteristics as those expected

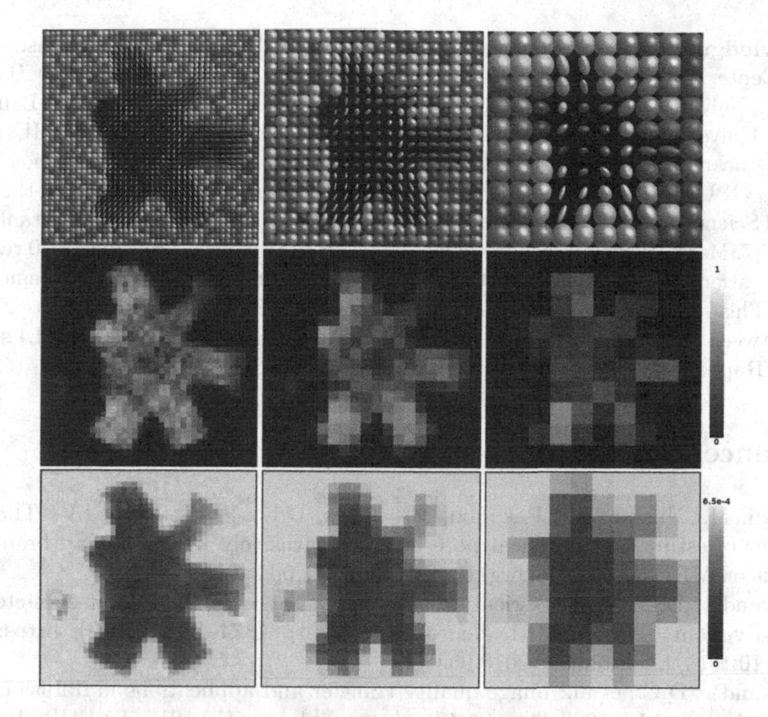

Fig. 6. Diffusion tensor estimated from the noiseless DW-MRI signal (top), fractional anisotropy map (middle) and mean diffusivity map (bottom). The mean diffusivity map is reported in $mm^2s^{-1}unit$. The voxel size of the image voxel size was set to 25 µm (left), 50 µm (center) and 100 µm (right) isotropic.

in real tissue. The end-to-end construction and simulation of each phantom was achievable in about one week, of which the substrate optimization procedure took about 5 to 6 days to complete. Finally, given the mesh information and the capability of handling the simulation independently for each of the three compartments, in the near future, we expect to be able to enhance the phantoms realism by including the transverse relaxation effects for each compartment individually. This will provide, for instance, an additional signal contrast to myelin and can be helpful for validating the biophysical modeling of the microstructural, including simulation-assisted machine learning approaches to it [22], and for the validation of methods that jointly use diffusion and relaxation information to detect and characterize pathology [11].

To summarize, we have shown an overview of the microstructural properties of the DiSCo dataset that are part of the MICCAI 2021 DiSCo Challenge. All of the computed maps, mesh information, and DWIs are to be made available publicly after the challenge event. We believe that these maps will boost the validation of connectomics and microstructure modeling. In addition, the phantoms can be reused to simulate more advanced protocols and even add new sources of contrast by tailoring the substrates and the biophysical properties to the specific research needs.

Acknowledgments. We acknowledge access to the facilities and expertise of the CIBM Center for Biomedical Imaging, supported by Lausanne University Hospital (CHUV), University of Lausanne (UNIL), Ecole polytechnique fédérale de Lausanne (EPFL), University of Geneva (UNIGE) and Geneva University Hospitals (HUG). We gratefully acknowledge the support of NVIDIA Corporation with the donation of the Titan Xp GPU used for this research. This project has received funding from the Swiss National Science Foundation under grant number 205320_175974 and Spark grant number 190297. Marco Pizzolato acknowledges the European Union's Horizon 2020 research and innovation programme under the Marie Skłodowska - Curie grant agreement No 754462. This research project is part of the MMINCARAV Inria associate team program between Empenn (Inria Rennes Bretagne Atlantique) and LTS5 (EPFL) started in 2019. Raphaël Truffet's PhD is partly funded by ENS Rennes.

References

1. Afzali, M., Nilsson, M., Palombo, M., Jones, D.K.: SPHERIOUSLY? The challenges of estimating spherical pore size non-invasively in the human brain from diffusion MRI. https://doi.org/10.1101/2020.11.06.371740
2. Alexander, D.C., et al.: Orientationally invariant indices of axon diameter and density from diffusion MRI. NeuroImage **52**(4), 1374–1389 (2010). https://doi.org/10.1016/j.neuroimage.2010.05.043
3. Alexander, D.C., et al.: Image quality transfer and applications in diffusion MRI. NeuroImage **152**, 283–298 (2017). https://doi.org/10.1016/J.NEUROIMAGE.2017.02.089
4. Andersson, M., et al.: Axon morphology is modulated by the local environment and impacts the noninvasive investigation of its structure-function relationship. Proc. Natl. Acad. Sci. USA **117**(52), 33649–33659 (2021). https://doi.org/10.1073/PNAS.2012533117
5. Basser, P.J., Mattiello, J., LeBihan, D.: MR diffusion tensor spectroscopy and imaging. Biophys. J. **66**(1), 259 (1994)
6. Callaghan, R., Alexander, D.C., Palombo, M., Zhang, H.: Config: contextual fibre growth to generate realistic axonal packing for diffusion MRI simulation. NeuroImage **220** (2020). https://doi.org/10.1016/j.neuroimage.2020.117107
7. Caruyer, E., Daducci, A., Descoteaux, M., Houde, J.c., Thiran, J.p., Verma, R.: Phantomas: a flexible software library to simulate diffusion MR phantoms. In: International Symposium on Magnetic Resonance in Medicine (ISMRM 2014), Milan, Italy (2014)
8. Close, T.G., Tournier, J.D., Calamante, F., Johnston, L.A., Mareels, I., Connelly, A.: A software tool to generate simulated white matter structures for the assessment of fibre-tracking algorithms. NeuroImage **47**(4), 1288–1300 (2009). https://doi.org/10.1016/J.NEUROIMAGE.2009.03.077
9. Coupé, P., Manjón, J.V., Chamberland, M., Descoteaux, M., Hiba, B.: Collaborative patch-based super-resolution for diffusion-weighted images. NeuroImage **83**, 245–261 (2013). https://doi.org/10.1016/j.neuroimage.2013.06.030
10. Daducci, A., Dal Palù, A., Lemkaddem, A., Thiran, J.P.: COMMIT: convex optimization modeling for microstructure informed tractography. IEEE Trans. Med. Imaging **34**(1), 246–257 (2015). https://doi.org/10.1109/TMI.2014.2352414
11. Fischi-Gomez, E., et al.: Multi-compartment diffusion MRI, T2 relaxometry and myelin water imaging as neuroimaging descriptors for anomalous tissue detection.

In: Proceedings - International Symposium on Biomedical Imaging, pp. 307–311, April 2021. https://doi.org/10.1109/ISBI48211.2021.9433856
12. Garyfallidis, E., et al.: Dipy, a library for the analysis of diffusion MRI data. Front. Neuroinform. 0(FEB), 8 (2014). https://doi.org/10.3389/FNINF.2014.00008
13. Ginsburger, K., Matuschke, F., Poupon, F., Mangin, J.F., Axer, M., Poupon, C.: MEDUSA: a GPU-based tool to create realistic phantoms of the brain microstructure using tiny spheres. NeuroImage **193**, 10–24 (2019). https://doi.org/10.1016/J.NEUROIMAGE.2019.02.055
14. Karunanithy, G., Wheeler, R.J., Tear, L.R., Farrer, N.J., Faulkner, S., Baldwin, A.J.: INDIANA: an in-cell diffusion method to characterize the size, abundance and permeability of cells. J. Magn. Resonan. **302**, 1–13 (2019). https://doi.org/10.1016/j.jmr.2018.12.001
15. Kumar, R., Vemuri, B.C., Wang, F., Syeda-Mahmood, T., Carney, P.R., Mareci, T.H.: Multi-fiber reconstruction from DW-MRI using a continuous mixture of hyperspherical von mises-fisher distributions. In: Prince, J.L., Pham, D.L., Myers, K.J. (eds.) IPMI 2009. LNCS, vol. 5636, pp. 139–150. Springer, Heidelberg (2009). https://doi.org/10.1007/978-3-642-02498-6_12
16. Lavdas, I., Behan, K.C., Papadaki, A., McRobbie, D.W., Aboagye, E.O.: A phantom for diffusion-weighted MRI (DW-MRI). J. Magn. Resonan. Imaging **38**(1), 173–179 (2013)
17. Neher, P.F., Laun, F.B., Stieltjes, B., Maier-Hein, K.H.: Fiberfox: facilitating the creation of realistic white matter software phantoms. Magn. Resonan. Med. **72**(5), 1460–1470 (2014). https://doi.org/10.1002/mrm.25045
18. Pautler, R.G., Silva, A.C., Koretsky, A.P.: In vivo neuronal tract tracing using manganese-enhanced magnetic resonance imaging. Magn. Resonan. Med. **40**(5), 740–748 (1998). https://doi.org/10.1002/mrm.1910400515
19. Penfield, W., Boldrey, E.: Somatic motor and sensory representation in the cerebral cortex of man as studied by electrical stimulation. Brain **60**(4), 389–443 (1937). https://doi.org/10.1093/BRAIN/60.4.389
20. Rafael-Patino, J., Girard, G., Truffet, R., Pizzolato, M., Caruyer, E., Thiran, J.P.: The diffusion-simulated connectivity (DiSCo) dataset. Data in Brief, July 2021
21. Rafael-Patino, J., Romascano, D., Ramirez-Manzanares, A., Canales-Rodríguez, E.J., Girard, G., Thiran, J.P.: Robust Monte-Carlo simulations in diffusion-MRI: effect of the substrate complexity and parameter choice on the reproducibility of results. Front. Neuroinform. **14**(8), 8 (2020). https://doi.org/10.3389/fninf.2020.00008
22. Rafael-Patino, J., et al.: DWI simulation-assisted machine learning models for microstructure estimation. Math. Visual. 125–134 (2020). https://doi.org/10.1007/978-3-030-52893-5_11
23. Ramirez-Manzanares, A., Rivera, M., Vemuri, B.C., Carney, P., Mareei, T.: Diffusion basis functions decomposition for estimating white matter intravoxel fiber geometry. IEEE Trans. Med. Imaging **26**(8), 1091–1102 (2007). https://doi.org/10.1109/TMI.2007.900461
24. Rensonnet, G., Rafael-Patiño, J., Macq, B., Thiran, J.P., Girard, G., Pizzolato, M.: A signal peak separation index for axisymmetric B-tensor encoding, October 2020. https://arxiv.org/abs/2010.08389
25. Romascano, D., et al.: HOTmix: characterizing hindered diffusion using a mixture of generalized higher order tensors (2019)
26. Tournier, J.D., Calamante, F., Connelly, A.: Robust determination of the fibre orientation distribution in diffusion MRI: non-negativity constrained super-resolved

spherical deconvolution. NeuroImage **35**(4), 1459–1472 (2007). https://doi.org/10.1016/j.neuroimage.2007.02.016
27. Tournier, J.D., Calamante, F., Connelly, A.: MRtrix: diffusion tractography in crossing fiber regions. Int. J. Imaging Syst. Technol. **22**(1), 53–66 (2012). https://doi.org/10.1002/IMA.22005
28. Tournier, J.D., et al.: Mrtrix3: a fast, flexible and open software framework for medical image processing and visualisation. NeuroImage **202**, 116137 (2019). https://doi.org/10.1016/j.neuroimage.2019.116137
29. Tristán-Vega, A., Aja-Fernández, S.: DWI filtering using joint information for DTI and HARDI. Med. Image Anal. **14**(2), 205–218 (2010). https://doi.org/10.1016/j.media.2009.11.001
30. Tuch, D.S.: Q-ball imaging. Magn. Resonan. Med. **52**(6), 1358–1372 (2004). https://doi.org/10.1002/MRM.20279
31. Yeh, C.H., Schmitt, B., Bihan, D.L., Li-Schlittgen, J.R., Lin, C.P., Poupon, C.: Diffusion microscopist simulator: a general Monte Carlo simulation system for diffusion magnetic resonance imaging. PLOS ONE **8**(10), e76626 (2013). https://doi.org/10.1371/JOURNAL.PONE.0076626
32. Zhang, H., Schneider, T., Wheeler-Kingshott, C.A., Alexander, D.C.: NODDI: practical in vivo neurite orientation dispersion and density imaging of the human brain. NeuroImage **61**(4), 1000–1016 (2012)

Author Index

Printed in the United States
by Baker & Taylor Publisher Services

Printed in the United States
by Baker & Taylor Publisher Services